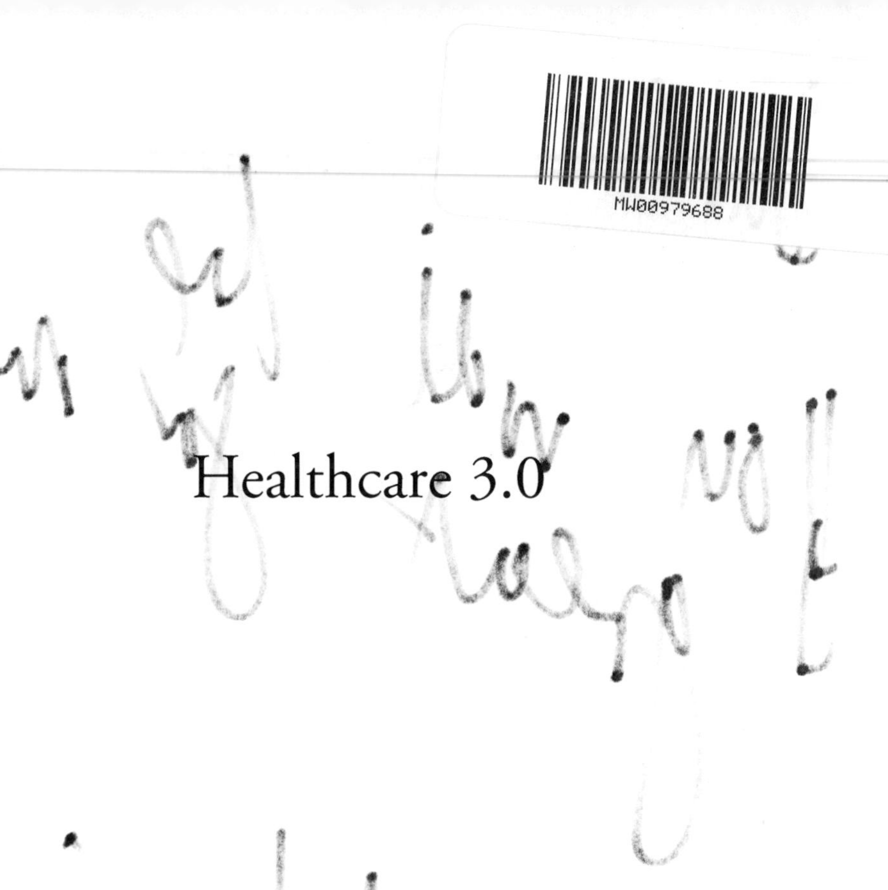

Healthcare 3.0

Darren:

You will be great for medicine

[illegible signature]

Healthcare 3.0

How Technology is Driving the Transition to Prosumers, Platforms and Outsurance

Rubin Pillay

Library of Congress Control Number:		2018911691
ISBN:	Hardcover	978-1-9845-5669-1
	Softcover	978-1-9845-5668-4
	eBook	978-1-9845-5667-7

Print information available on the last page.

Rev. date: 10/29/2018

To order additional copies of this book, contact:
Xlibris
1-888-795-4274
www.Xlibris.com
Orders@Xlibris.com
770707

CONTENTS

CHAPTER 1

Introduction

Over the coming decades, humanity will encounter some of the greatest transitions any generation has ever had to face. Technological disruption is reshaping every part of our lives—every business, every industry, every society, even what it means to be "human." Exponential technologies are on the cusp of solving some of humanity's biggest challenges, and health care is set to be one of the chief beneficiaries. We are already in the midst of this medical revolution driven by the convergence of exponential hardware, software, communication, and biomedical technologies. This book outlines how this convergence is set to transform the three key pillars of health care—patients, providers, and payers—and, for the very first time, force a synergism that's going to help us solve the health-care crisis.

It's no secret that the world of health care has changed dramatically for the better over the last few centuries. During the rough and tough Stone Ages, people only lived twenty years. This improved

30 percent to twenty-six years in the Bronze and Iron Ages. By the middle ages, we made it to forty years, and our push toward longevity only really began during the industrial revolution with the advent of modern medicine. Commensurate with the increase in our life spans, we also saw a dramatic improvement in other measures, which reflect a nation's health outcomes. Since 1900, infant mortality has decreased 90 percent, and maternal mortality has decreased 99 percent.

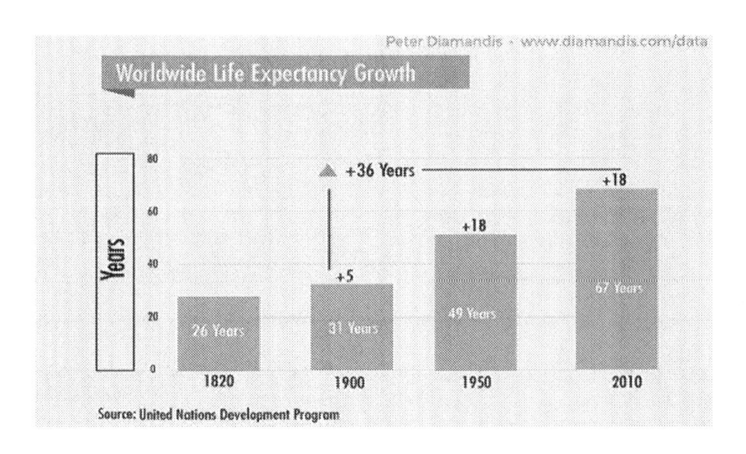

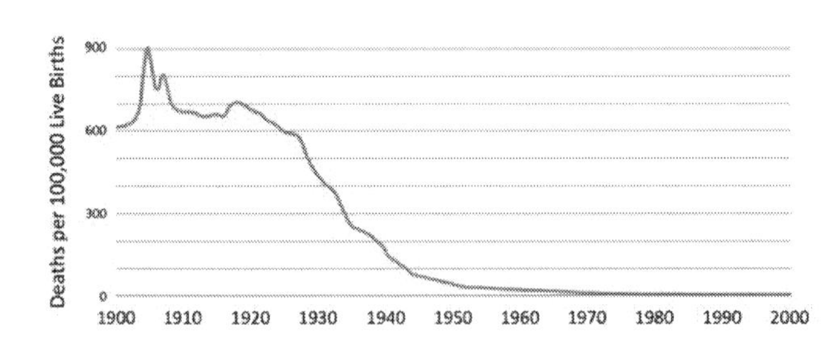

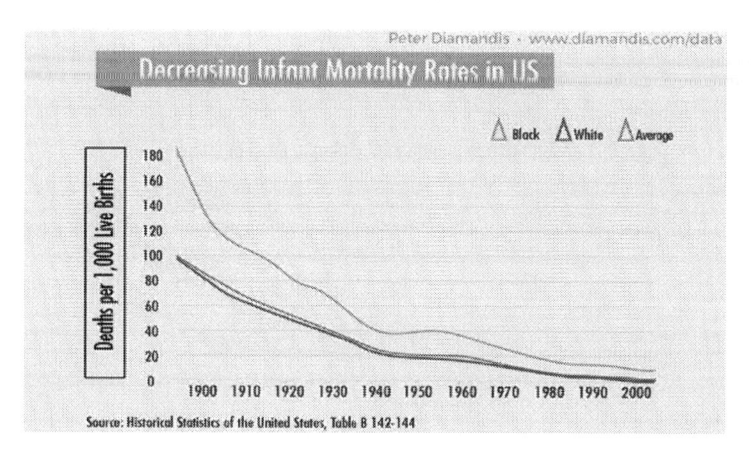

Upon close reflection, I have observed three key periods and ways that health care has changed—and continues to change—in the last century. The century predating the mid-nineties saw the most significant improvements in health outcomes. Since 1900, the average life span of people in the United States has lengthened by greater than thirty years; twenty-five years of this gain are attributable to advances in public health—*health care 1.0.* Many notable public health achievements have occurred during the early 1900s. Control of infectious diseases has resulted from clean water and improved sanitation. Infections such as typhoid and cholera transmitted by contaminated water, a major cause of illness and death early in the twentieth century, have been reduced dramatically by improved sanitation. In addition, the roll out of primary preventive measures such as large-scale vaccinations has been critical to successful public health efforts to control infections such as tuberculosis. The list of achievements that highlight the contributions of public health and the impact of these contributions on the health and well-being of people across the globe has continued to grow to this day, and interventions in the areas of tobacco use, motor-vehicle safety, risk factor modification, nutrition, family planning, and water fluorination continue to significantly impact health outcomes positively.

The advent of modern medicine characterized by a biomedical scientific agenda and the modern hospital in the mid- twentieth century skyrocketed our mean life expectancy to close to seventy years and further dropped infant and maternal mortality rates, at least in the developed world. In 1945, the president's science advisor Vannevar Bush wrote in *Science, the Endless Frontier* that basic scientific research was "the pacemaker of technological progress" and that "new products and new processes do not appear full-grown. They are founded on new principles and new conceptions, which in turn are painstakingly developed by research in the purest realms of science." He recommended the creation of what would become the National Institutes of Health (NIH) that was created in 1948 and the National Science Foundation created in 1950. Since 1945, biomedical research, the primary source of the world's new drugs and medical devices, has been viewed as the essential contributor to improving the health of individuals and populations in both the developed and developing world—*health care 2.0.*

The great thing about medical research is that it will never abate. There will always be new discoveries and innovation, new ways to improve health and health care. Thousands are working in labs right now, all on the cusp of a new discovery. By 2025 we will be doubling medical knowledge every seventy three days! If we think that we are doing well now, imagine what these discoveries will do for us in the future. Medical research, like public health, will continue to hold promise for the future.

Despite these significant advances in the biomedical sciences and health care over the last century, we are being barraged with negative news. This constant onslaught often distorts our perspective on the future. But the good news is that we are on the cusp of radical change. We are destined for a future of near perfect health, a future that will be the result of the exponential growth of advanced technologies such as digital medicine, artificial intelligence, gene editing, nanotechnology, robotics—the list goes on—and their convergence, *health care 3.0*. This technological storm will transform medicine and health care in ways that sounded like science fiction a mere decade ago. It will give medical professionals, patients, and key industry players the unprecedented ability to make appropriate health care more accessible, affordable, and humanistic and will propel three major trends that will see health care flipped on its head: the transformation from patient to prosumer, the move from a pipeline-based approach to sick-care delivery to a platform-based approach, and a shift from insurance-based payer approach to an outsurance-based approach.

Health Care Today

Although public health, medical science, and technology have advanced at an unprecedented rate during the past century, health care in the USA and across the globe is in crisis. We are plagued with erratic quality, unequal access, and sky-high costs. The Institute of Medicine describes the difference between the health care that we now have and the health care that we could have as a chasm and not just a gap.

A number of supply side and demand side factors have combined to create this chasm. At the heart of the problem has been our inability to translate knowledge into practice and to apply new technology safely and appropriately. This has been driven by the increased complexity of health care, resulting primarily from the massive expansion of our knowledge and technology base—more to know, more to do, more to manage, more to watch, and more people involved than ever before. Consequently, the health-care delivery system is poorly organized to meet the challenges at hand. The delivery of care often is overly complex and uncoordinated, leading to decreased patient safety, wasted resources, poor access, and an inability to capitalize on the strengths of all health professionals involved to ensure that care is appropriate, timely, safe, and cost effective. If the system cannot consistently deliver today's science and technology, it is even less prepared to respond to the extraordinary advances that are on the horizon.

On the demand side, health-care needs have changed as well. Population growth coupled with people living longer has resulted in a real and relative increase in demand for health-care services. The increase demand is for chronic and noncommunicable conditions, such as heart disease, diabetes, chronic obstructive airways disease and cancer, while the system is structured more to deliver acute episodic care. This is compounded by the fact that more than 40 percent of people with chronic conditions have more than one such condition, which makes a compelling case for more sophisticated mechanisms to coordinate care. Our inability to do this is reflected in the fact that on average, patients received 54.9 percent of recommended care with little difference among the proportion of recommended preventive care provided (54.9 percent), the proportion of recommended acute care provided (53.5 percent), and the proportion of recommended care provided for chronic conditions (56.1 percent). These deficits in adherence to recommended processes for basic care pose serious threats to the health of the American public.

At the macro level, the United States spends far more on health care on a per capita basis and as a percentage of GDP than other high-income countries, with spending levels that rose continuously over the past three decades. At current rates, by 2025, nearly one-fourth of the US GDP will be spent on health care. However, when compared to ten other OECD countries, the United States ranks last in health-care system performance; last in access, equity, and health-care outcomes; and next to last in administrative efficiency as reported by patients and providers. Only in care process does the United States perform better, ranking fifth among the eleven countries.

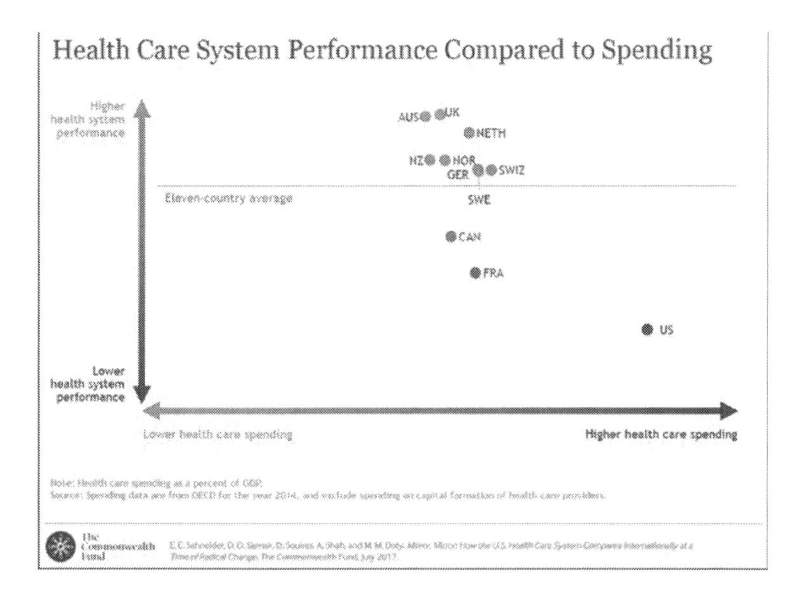

At the micro level—and this is scary—the Institute of Medicine asserts that diagnostic errors occur daily in every health-care setting nationwide and estimates that most people will experience at least one diagnostic error in their lifetime, sometimes with devastating consequences. In addition are the following:

- About 5 percent of adults who seek outpatient care annually suffer a delayed or wrong diagnosis.
- Postmortem research suggests that diagnostic errors are implicated in one of every ten patient deaths.
- Chart reviews indicate that diagnostic errors account for up to 17 percent of hospital adverse events.
- Diagnostic errors are the principle cause of paid malpractice claims and are almost twice more likely to end in a patient's death than claims for other medical mishaps. They also represent the biggest share of total payments.

- On average, a hospital patient is subject to at least one medication error per day.

More than 250,000 Americans die each year from medical errors, making it the third leading cause of death in the United States.

So despite spending nearly twice as much as several other countries, the United States' performance is lackluster at best. To gain more than incremental improvement, however, we must pursue different approaches. Technology has transformed almost every aspect of our lives—mostly in a positive way—and is set to transform health care for both health-care workers and patients across the world. It has the potential to transform unsustainable health-care systems into sustainable ones; equalize the relationship between medical professionals and patients; provide cheaper, faster, and more effective solutions for diseases. Technologies could win the battle for us against cardiac disease, cancer, diabetes, and other chronic conditions and could simply lead to healthier individuals living in healthier communities.

CHAPTER 2

Key Technology Drivers

We live in unbelievable times—instant access to information and global news, love just a click away, entertainment 24/7 on always connected devices, banking and travel bookings from home. Powerful technologies that were once only available to huge organizations and governments are becoming smaller, cheaper, faster, better, and more accessible to every single one of us, and this radical transformation is set to redefine our very notion of health and health care.

The disruption and transformation of medicine has already begun, and we will witness medicine advance more in the next decade than it has in the last century. We're in the midst of an avalanche of converging technologies in medical science, software, hardware, and communications, and this perfect storm is transforming medicine and health care in ways that sounded like science fiction a mere decade ago . . . and it's happening fast.

No doubt you've heard of Moore's law. It refers to the exponential price-performance improvements of integrated circuits. Moore observed that the number of transistors in a dense integrated circuit doubles approximately every two years. From two transistors in 1958 to 7.1 billion in 2012—a hundred billion-fold improvement in forty years. The cost of transistors also halved year on year—from $1 in 1971 to $0.0000001 in 2012.

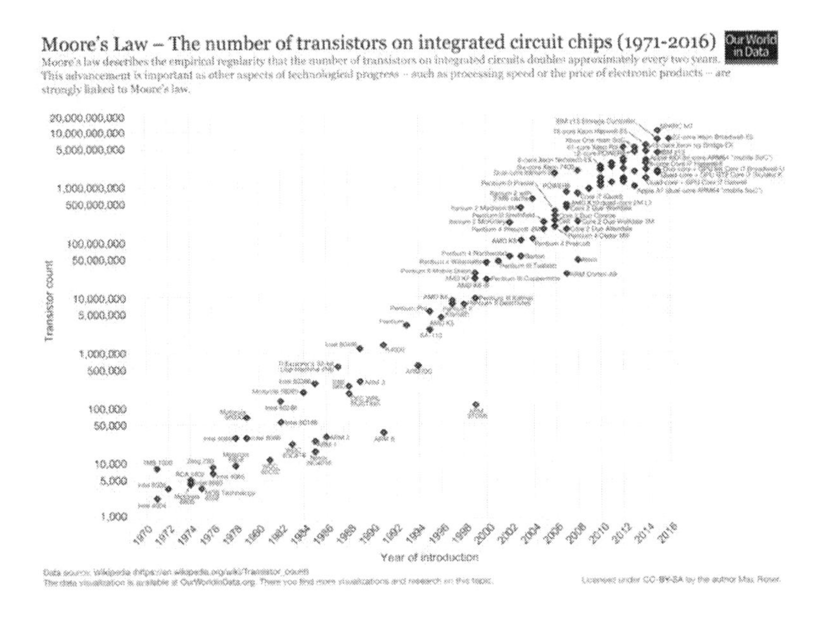

The concept of exponentially is not unique to computing. For a technology to be "exponential," the capability doubles each year, or the cost drops by half. Take the case of human genome sequencing. The time to read a human genome has sped up by a factor of at least a thousand and dropped in price over a factor of ten thousand since 2001. Yes, ten-thousand-fold improvement in cost alone. For what it cost to read the first human genome, we could read the entire DNA of ten thousand people. The first attempt to read the human genome began in 1985, and it took fifteen years. The total cost was $3 billion. The second attempt that started in 2000 took only three years and cost a mere $100 million. Today we can read an entire human genome in a day for a thousand dollars, outpacing Moore's law since January 2008.

In fact today, companies like Nebula Genomics (https://www.nebulagenomics.io/) will actually pay you to sequence your genome.

Although the current pace of progress warrants an exponential mind-set, we are not by nature equipped to process exponential growth. Our intuition is to use our assessment of how much change we've seen in the past to predict how much change we'll see going forward—linear thinking. Thinking exponentially though is key to discovering potential new opportunities and building innovative solutions. To get a gut feeling for the difference between linear thinking and exponential thinking, let's look at the difference between taking thirty linear steps and thirty exponential steps. The former will get you thirty paces ahead, while the latter will get you twenty-six times around the earth. Clearly, thinking linearly can prove costly for businesses, governments, and individuals alike. If we can better plan for the accelerating pace, we can ease the transition from one paradigm to the next and better ensure our success in the future.

In this chapter, we explore some of the exponential technologies that are set to impact health care. The solutions to the world's most pressing health and health-care challenges lie in the intersection of these exponential technologies. When two or more of these technologies are used in combination—converging exponential technologies—we will begin to unfathomably unlock productive capabilities and begin to understand how to solve the world's most challenging problems. Exponential technologies will drive less expensive, more efficient, and more accessible care delivery on a global scale. There has never been a better time to resolve the cost, quality, and access challenges that have plagued health care for the last few decades.

Artificial Intelligence (AI)

The AI revolution is the most profound transformation human civilization will experience in all history. This is because this revolution is characterized by being able to replicate human intelligence that is arguably the most important and powerful attribute of human civilization. Strong AI—and we are not quite there yet—will be as versatile as humans when it comes to solving problems. And according to futurist Ray Kurzweil, even AI that can function at the level of human intelligence will still outperform humans because of several aspects unique to machines:

- Machines can pool resources in ways that humans cannot.
- Machines have exacting memories.
- Machines "can consistently perform at peak levels and can combine peak skills."

The term "artificial intelligence" was coined in 1956 and refers to the reproduction of human cognitive functions—such as problem solving, reasoning, understanding, recognition—by artificial means, specifically by computer. Most of us are barely aware of it, but AI is already part of our lives—it's in our cars telling us when it's time for the engine to be serviced based on driving patterns, it's in our Google searches and suggestions from Amazon, it's the chatbot at the other end at a "call center." It's pretty much becoming an integral part of our lives. Apple's Siri processes about two billion requests a week. One of AI's biggest potential benefits is to help people stay healthy.

AI is getting increasingly sophisticated at what humans do, but more efficiently, more quickly, and at lower cost. The potential in health care is vast and is on the rise. It is solving a variety of problems for patients, physicians, hospitals, and the health-care industry overall. AI will help us stay healthy, so we don't need a doctor, or at least not as often, by enabling us to stay well, detect disease early, improve diagnostic capability, and aid decision-making, treatment, and prognostication.

Companies such as in Fitbit (https://www.fitbit.com/home), Apple (https://www.apple.com), and Samsung (https://www.samsung.com/us/) offer consumers devices such as smart watches and activity trackers linked to their operating system, which enables people to track their fitness. These applications and others all encourage healthier behavior in individuals and help with the proactive management of a healthy lifestyle. It puts consumers in control of health and well-being. Additionally, AI increases the ability for health-care professionals to better understand the day-to-day patterns and needs of the people they care for, and with that understanding, they are able to provide better feedback, guidance, and support.

IBM's Watson AI (https://www.ibm.com/watson/) technology is already tackling a wide range of the world's biggest health-care challenges, including cancer, diabetes, and drug discovery. In oncology, Watson is at work supporting cancer care in more than 150 hospitals and health organizations in eleven countries, and a large, growing body of evidence supports the use of Watson in health care.

Companies such as Enlitic (https://www.enlitic.com/) also use deep learning to make doctors faster and more accurate by incorporating a wide range of unstructured medical data, including radiology and pathology images; laboratory results, such as blood tests and EKGs; genomics; patient histories; and electronic health records (EHRs) to enable higher accuracy and deeper insights for every patient.

Similarly, AI is already being used to detect diseases, such as cancer, more accurately in their early stages. For example, according to the American Cancer Society, 12.1 million mammograms are performed annually in the United States, but a high proportion of these mammograms yield false results, leading to one in two healthy women being told they have cancer. The use of AI is enabling review and translation of mammograms thirty times faster, with 99 percent accuracy, reducing the need for unnecessary biopsies as well as reducing the uncertainty and stress of a misdiagnosis. The accuracy gap between the human and digital eye is expected to widen further and soon. As machines become more powerful and deep-learning approaches gain traction, they will continue to advance such diagnostic fields as radiology (CT, MRI, and mammography interpretation), pathology (microscopic and cytological diagnoses), dermatology (rash identification and pigmented lesion evaluation for potential melanoma), and ophthalmology (retinal vessel examination to predict the risk for diabetic retinopathy and cardiovascular disease). The NEJM predicts that AI will displace much of the work of radiologists, dermatologists, and anatomical pathologists. These physicians focus largely on interpreting digitized images, which can easily be fed directly to algorithms instead.

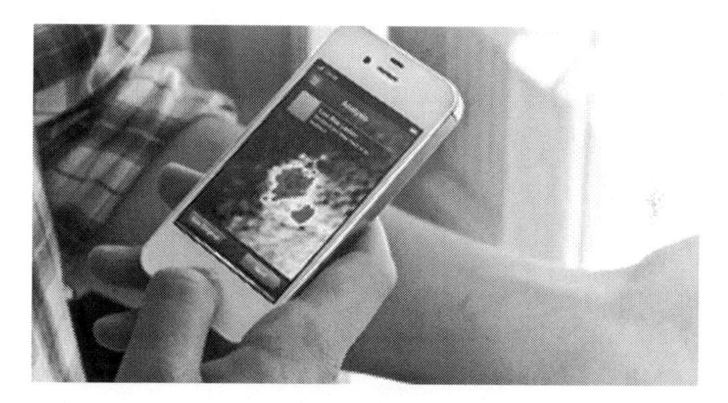

Skinvision

Already new research is suggesting that artificial intelligence may be better than highly-trained humans at detecting skin cancer. A study conducted by an international team of researchers pitted experienced dermatologists against a machine learning system, known as a deep learning convolutional neural network, or

CNN, to see which was more effective at detecting malignant melanomas. The results? "Most dermatologists were outperformed by the CNN," the researchers wrote in their report, published in the journal Annals of Oncology.

AI technology like SkinVison (https://www.skinvision.com/) target the early early detection of skin cancer which is the most common cancer in Western countries. SkinVision offers a complete skin cancer detection service that combines a clinically-proven machine-learning technology with the knowledge of in-house dermatologists. The App and cloud based system makes it possible to detect skin cancer at an early stage when it's most treatable and has less expensive treatment options. As a result, the early detection of skin cancer allows you to save on medical costs occurring from future treatment.

Research published in the Journal of the American Medical Association also found that in the setting of a challenge competition, some deep learning algorithms achieved better diagnostic performance than a panel of 11 pathologists participating in a simulation exercise designed to mimic routine pathology workflow; algorithm performance was comparable with an expert pathologist interpreting whole-slide images without time constraints.

Scientists from Google and its health-tech subsidiary Verily have discovered a new way to assess a person's risk of heart disease using machine learning. By analyzing scans of the back of a patient's eye, the company's software is able to accurately deduce data, including an individual's age, blood pressure, and whether or not they smoke. This can then be used to predict their risk of suffering a major cardiac event — such as a heart attack — with roughly the same accuracy as current leading methods. The algorithm potentially makes it quicker and easier for doctors to analyze a patient's cardiovascular risk, as it doesn't require a blood test. But, the method will need to be tested more thoroughly before it can be used in a clinical setting.

Babylon (https://www.babylonhealth.com), a UK-based digital health company, launched an application that offers medical AI consultation based on personal medical history and common medical knowledge. Users report the symptoms of their illness to the app that checks them against a database of diseases using speech recognition. After taking into account the patient's history and circumstances, Babylon offers an appropriate course of action. The app will also remind patients to take their medication and follow up to find out how they're feeling. Through such solutions, the efficiency of diagnosing patients can increase by multiple times while the waiting time in front of doctor's examining rooms could drop significantly. Babylon recently partnered with the government of Rwanda—one of the world's poorest countries—to make free medical advice and services available to its population.

A host of other AI products are aimed at supporting clinical decision-making:

- Doc.ai (https://doc.ai) plans to use AI to interpret lab results. The company's first product will interpret blood tests, genetics tests, and gradually add other kinds of tests.

- Careskore (https://www.careskore.com) basically predicts how likely a patient will be readmitted to a hospital through its Zeus algorithm in real time. This information is based on a combination of clinical, labs, and demographic and behavioral data.
- Oncora Medical (https://oncoramedical.com) uses a data analytics platform to help doctors design personalized radiation treatment plans for patients.
- Sentrian (http://www.sentrian.com) smart algorithms tell people they are going to be sick, even before they experience symptoms.
- Medical appointment booking app Zocdoc (https://www.zocdoc.com) has launched Insurance Checker, a new feature powered by artificial intelligence seeking to ease a process pain point common for both patients and providers. Insurance Checker targets deciphering, understanding, and verifying health insurance.

Artificial intelligence is also set to dramatically improve the drug discovery process that is long and costly. Atomwise (https://www.atomwise.com) uses deep learning algorithms to analyze molecules and predict how they might act in the human body—including their potential efficacy as medication, toxicity, and side effects—at an earlier stage than in the traditional drug discovery process. During the recent Ebola outbreak, two drugs predicted by Atomwise's artificial intelligence technology were proven to significantly reduce Ebola infectivity. These drugs were intended for unrelated illnesses and their potential to treat Ebola was previously unknown. Atomwise accomplished this in twenty-four hours. There are currently about one hundred startups using AI in drug discovery.

Robotics

While there are concerns for the automation of the medical workforce, the benefits of robotics in health care cannot be ignored. The precision, accuracy, and mobility of medical robotics will allow us to serve more humans around the world faster and cheaper. Robots are taking on more challenging tasks in today's medical applications with their improved capabilities. Robotic-assisted surgical procedures, robotic medicine dispensing, rehabilitation and movement therapy where robotics assist the patient, and assistance for the elderly or individuals with disabilities in performing daily activities of living are benefitting from the latest robot technologies. They are improving productivity during procedures, enabling users to perform tasks that are more challenging, improve patient comfort, and reduce costs for medical staff and, ultimately, patients. Robotics technology can also incorporate telepresence and, therefore, becomes an important tool in providing health care in rural and remote locations.

Intuitive's Da Vinci Surgical System (https://www.intuitivesurgical.com/) was one of the first to enable surgeons to operate with enhanced vision, precision, and control. With over three million surgeries worldwide to date, the surgeon is always 100 percent in control at all times using 3-D HD vision inside the body, with precise movements that don't have the tremors of a human hand.

Medrobotics's Flex Robotic System (https://medrobotics.com/gateway/flex-system-int/) is a robot-assisted surgical platform with flexible robotic scopes, flexible instruments, and HD visualization and was designed to efficiently integrate into an operating room. Medrobotics received FDA clearance for Flex for head and neck procedures in July 2015 and colorectal procedures in May 2017. This system translates the surgeon's hand motions into precise movements that let them access difficult-to-reach anatomy. The Flex Robotic System is able to do natural orifice surgeries in the colon and at the top of the voice box, which are hard to reach.

The small mobile platform of the Flex can be wheeled into a new room and set up or reconfigured in five to six minutes. No special rooms are needed for robotic surgeries. Cost savings may come from shorter hospital stays, from reduced infection rates and other complications. Its efficiency also allows hospitals to perform more procedures per operating room since setup times are short and the robot can move from room to room "just in time" for the next procedure.

The next generation of precision surgery robots will be fully autonomous. In 2017, a robot carried out a dental operation without help from humans for the first time, carrying out implant surgery on a patient in China. It followed a set of preprogrammed commands to fit the implants into the patient's mouth but was able to make adjustments as the woman moved. Although medical staff were present during the one-hour surgery in Xi'an, Shaanxi Province, they did not play an active role.

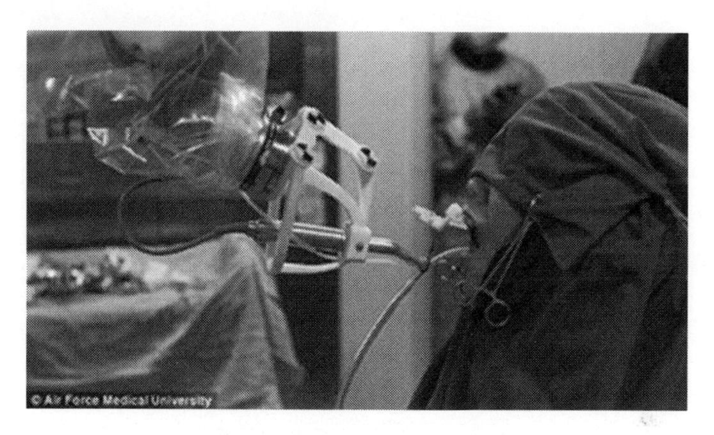

Robotic implant surgery

During a hospital stay, patients interact with nurses the most. They draw blood, check your vital signs, check on your condition, and take care of your hygiene if needed. They are often overwhelmed by physically and mentally daunting tasks, and the result is often an unpleasant experience for everyone involved. Robotic nurses will help carry this burden in the future. They are designed to be able to carry out repetitive tasks. This way, the staff has more energy to deal with issues that require human decision-making skills and empathy. Certain robots can even take your blood sample (https://www.youtube.com/watch?v=IpdTeGPruFA).

But robotics in health care is so much more than drawing blood. With a remote-controlled robot such as the ones developed by Anybots Inc. (http://www.anybots.com/), caretakers can interact with their patients, check on their living conditions and the need for further appointments. This would help efficiency a great deal by eliminating the time-consuming home visits. Companies producing and the ones maintaining the system will have to make great efforts to alleviate privacy concerns. As with every such device, it must be near impossible to access for nonauthorized personnel. With the proper safeguards in place, these robots can greatly improve the lives of caretakers and patients alike. Telepresence technologies such as Suitable Technology's BEAM (https://suitabletech.com/products/beam) and InTouch Health (https://intouchhealth.com/) also allows physicians to beam into locations around the world for consultation and rounds at hospitals as well as provide health care in rural and remote locations.

Bionics, exoskeletons, and next-generation wearable robots are set to revolutionize how we treat and care for those with diminished functionality. With the help of these devices, paralyzed people can walk, and patients with stroke or spinal cord injury can be rehabilitated. They can enhance strength in order to allow a nurse to lift an elderly patient. While they have many exciting uses, it's important to remember that currently, they are costly to make and power, so at least, at first, they will not be available for everyone. However, in some cases, insurance companies had to cover the costs. Because of this, it has the potential to deepen already existing social and economic inequalities. Decision makers have to lay the groundwork to regulate the use of such devices. They will have to stay up to date on their capabilities to prevent misuse.

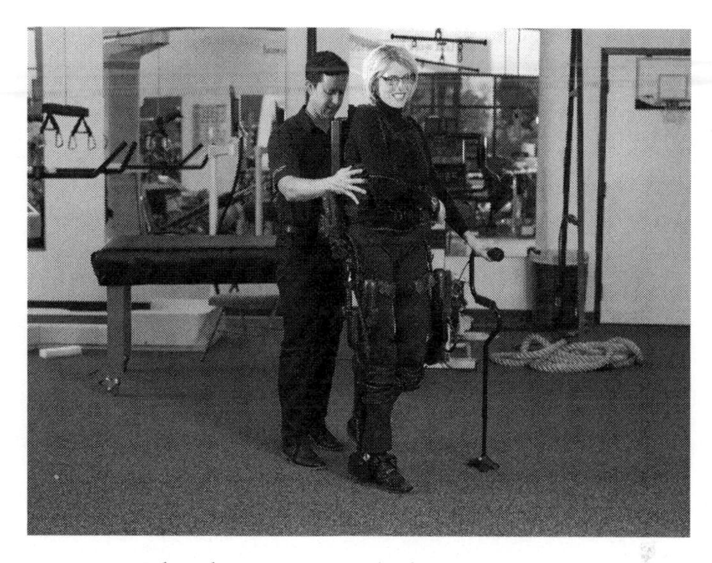

Eksobionics exoskeleton system

Robots are also now working in a hospital and can replace the jobs of four people each. Eight of the robots called Noah (https://www.youtube.com/watch?v=VMUaUDHHbFs) are working in a busy hospital where they run errands for nurses. Each of the little bots can carry documents and medicine weighing up to 330 kilograms, but bosses at the hospital believe they are capable of much more. The Noahs that are based at the Guangzhou Women and Children Medical Center in the capital of South China's Guangdong Province currently operate between the pharmacy and the nurse's station. Using GPS technology, a Noah bot is able to navigate the hospital's complex wings with ease and can carry ten times as much as one human staff member.

The hospital estimates the machines will save nurses up to 1,207 miles of walking every year. The intelligent Noah bots are programmed with phrases such as "Here I go," "I'm entering the lift," and "I've been obstructed" to notify their human colleagues throughout the day.

Similarly, Xenex (https://www.xenex.com/), a Texas-based company produces a unique robot. It uses high intensity ultraviolet light to quickly and efficiently disinfect any space in a health-care facility. The Xenex robot is more effective in causing cellular damage to microorganisms than other devices designed for disinfection. It reduces the number of hospital-acquired infections. It's yet another example of how robotics in health care helps hospital staff to decrease workload and will lead to a much friendlier environment.

Xenex

The Origami robot developed by researchers at MIT, the University of Sheffield, and the Tokyo Institute of Technology, despite its size (less than one centimeter), is just as impressive as a super strong carrier one. When swallowed, the capsule containing it dissolves in the patient's stomach and unfolds itself. Controlled by a technician with the help of magnetic fields, it can patch up wounds in the stomach lining or safely remove foreign items, such as swallowed toys.

Change is often scary, and robotics in health care is a big one. It has the potential to do so much good: to bring medical care to regions where even today there is none to be found, to make the production and distribution of pharmaceuticals cheaper and more efficient, to lighten the load of medical professionals, and to help people walk again. To reap the benefits and avoid the potential dangers of such a technological revolution, we need to keep informed about the strides that science makes so that we can better prepare and adapt to the not-so-distant future where robots play a crucial role and work closely with us.

IoT, Biosensors and Trackers

The sensor revolution is part of the so-called Internet of Things (IoT). The IoT has been defined in a variety of ways, but it is generally described as an ecosystem of technologies that monitor the status of objects, individuals, and environments, capturing meaningful data, potentially interacting with one another, and communicating valuable information over networks to software applications that can accurately and quickly analyze the data to glean important information and trends. The IoT promises to be the most disruptive technology since the World Wide Web. Nowhere does the IoT offer more promise than in health care where it is already being applied to improve care quality, access, and costs.

These technologies are being used by leading-edge organizations to grapple with both operational and clinical problems in both inpatient and outpatient environments. Coupled with radio frequency identification (RFID) technologies, IoT can help hospitals better manage demand, inventories, assets, and throughput as well as realize better integration of services and processes. Clinically, the IoT offers improved monitoring and managing of vital signs, infusions, medical device interoperability, medication administration, potential harmful events, and surgeries, to name a few.

Sensors and wearables that monitor physiological data of older people and individuals with chronic conditions can facilitate timely clinical interventions. This include wearables such as the Vital Connect (https://vitalconnect.com/) that can transmit your electrocardiogram data, vital signs, posture, and stress levels anywhere on the planet. Some consumer-focused health and fitness sensors such as Fitbit are widely used by enthusiastic individuals to gather quantified data about their health.

Many sensors and wearables have been developed to detect falls, epileptic seizures [Embrace by Empatica (https://www.empatica.com/)], and heart attacks [Kardia Mobile (https://www.alivecor.com/) and WIWE (http://shop.mywiwe.com/en/)] in older people and susceptible individuals and then send alarm signals to caregivers or emergency response teams.

Sensors are also used to monitor, assess, and improve patient compliance. In 2017, the Food and Drug Administration approved the first digital pill for the United States that tracks if patients have taken their medication. The pill called Abilify MyCite (https://www.abilifymycite.com/) is fitted with a tiny ingestible sensor that communicates with a patch worn by the patient. The patch then transmits medication data to a smartphone app that the patient can voluntarily upload to a database for their doctor and other authorized people to see. Abilify is a drug that treats schizophrenia, bipolar disorder, and is an add-on treatment for depression.

A range of physiological and biological sensors is also enabling the early detection of and monitoring of diseases. Google is developing a full range of internal and external sensors (such as Google's smart contact lens) that can monitor the wearer's vitals ranging from blood sugar levels to blood chemistries, while Profusa Inc. (https://profusa.com/), a San Francisco Bay Area–based life science company is developing a family of tiny biosensors composed of a tissue-like hydrogel, similar to a soft contact lens, that are painlessly placed under the skin with a single injection. Smaller than a grain of rice, each biosensor is a flexible fiber about five millimeters long and half a millimeter wide that continuously monitor the concentration of a body chemical, such as oxygen, glucose, or other biomolecule of interest.

Glaucoma typically occurs when a blockage in the eye's drainage channels causes aqueous humor fluid to accumulate within the eye faster than it can drain out. This leads to an increase in intraocular pressure, which can in turn damage the optic nerve, resulting in permanent loss of vision. It's therefore important for patients to have that pressure checked regularly, so that any significant increases can be addressed via measures such as changes in medication, or even surgery. Unfortunately, though, these pressure-checks have to be performed at an ophthalmologist's office, and thus usually only take place a few times a year. Any fluctuations in intraocular pressure that occur between those visits will simply be missed.

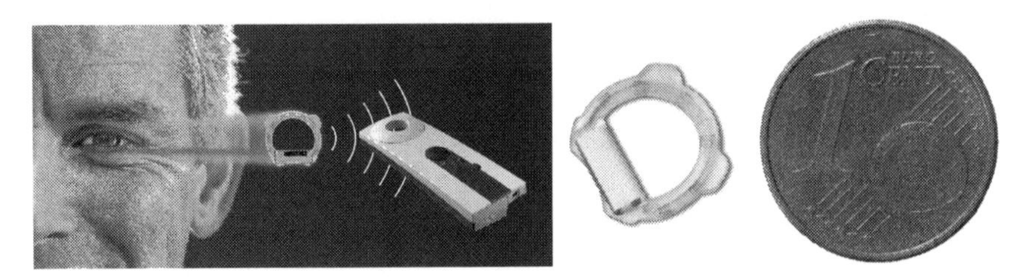

EYEMATE

It was with this problem in mind that Eyemate was created (http://www.my-eyemate.com/en/). Designed via a collaboration between Germany's Implandata Ophthalmic Products and the Fraunhofer Institute for Microelectronic Circuits and Systems, the thin and flexible ring-shaped implant can be placed behind the eye's iris during standard cataract surgery – it goes in through the same incision that is used to remove the eye's clouded natural lens, and to insert its artificial replacement. This is done using a handheld reading device which is held up to the eye without contacting it, temporarily powering the implant by delivering a 2-second weak magnetic pulse. The implant proceeds to measure the intraocular pressure and temperature, and transmits that data back to the reader. Users see the numerical reading on the reader's digital display, plus they're able to upload it to a secure online database that can be accessed anytime by their physician. Additionally, they can use an app to track their readings over time. Patients are subsequently able to perform pressure readings as often as they wish, within their own home.

By combining physiological sensors with activity monitors and consumer-end electronic devices, this application of digital health can be used for early detection of symptoms and adverse changes in a patient's health status, facilitating timely medical interventions. All this will result in making vast amounts of new health and wellness data available to health-care providers. This will further fuel the need for health systems to implement robust analytical and data-driven improvement systems that allow providers to optimally manage the health and well-being of populations.

The much-anticipated sensor revolution in health care is underway. And this is just the beginning. The pace of change in this area will almost certainly experience exponential growth as new use cases and technologies continue to emerge to address the need for more affordable, accessible, high-quality, and patient-centric care.

3D Printing

Three-dimensional printing involves producing several successive layers atop one another, ultimately producing a 3D object. Another term used is "additive manufacturing," which may be more accurate in describing the process. It uses a range of materials including various plastics and metals as well as biological material, namely, cells. The technology for 3D printing has evolved since its invention in the 1980s to the point that it is now economical to use in small-scale production and for customized purposes. While the technology was initially used in engineering for the production of prototypes, there are several applications for 3D printing in medicine.

It is recognized that medical uses for 3D printing, both actual and potential, will bring revolutionary changes. They can be organized into several broad categories, including creation of customized prosthetics, implants, and anatomical models; tissue and organ fabrication; manufacturing of specialty surgical instruments; pharmaceutical research regarding drug fabrication, dosage forms, delivery, and discovery; and manufacturing of medical devices. Benefits provided by application of 3D printing in medicine include not only the customization and personalization of medical products, drugs, and equipment but also cost-effectiveness, increased productivity, the democratization of design and manufacturing, and enhanced collaboration.

The technology is low cost and can be carried out in remote locations with limited resources. The cost has progressively dropped to the point that tabletop printers are common, and their cost has continued to decrease over the years to a range of $300–$3,000. In addition, the cost of the plastic used as feedstock, or the printing material, is just pennies per gram.

Three-dimensional printing is ideally suited to both on-demand and custom device production. On-demand manufacturing will make medical devices cheaper and more readily accessible to millions, and it will make scarce resources such as organs-for-transplant abundantly available. In general, the applications of

3D printing in medicine are diverse, and in some areas, it has or may disrupt the whole market. Examples include the following:

- *Medical devices.* The number of 3D-printed medical devices is enormous and is growing steadily. For example, 3D Systems (https://www.3dsystems.com/industries/healthcare) is 3D-printing precise dental and anatomical models (tumor and organ), custom surgical guides, implantable devices, exoskeletons, hearing aids, prosthetics, and braces for scoliosis and other applications. They include an inexpensive high-quality stethoscope developed for poor hospitals in the Gaza Strip, a 3D-printed winch used in endogenous laser therapy (EVLT) for the treatment of varicose vein removal that is presented by the Polish company Zortrax, a multisensory perception simulator (SPPS) device developed by the Center for Hearing and Speech in Kajetany, Poland, glass frames you can order or 3D print yourself or 3D-printed lenses. The latter were a special challenge in view of the layered structure of the 3D printouts, but the glass barrier has been broken. The marketing of the 3D-printed titanium device fast-forward bone tether plate (https://www.medshape.com) that facilitates less-invasive foot surgery and eliminates the need to drill through the bone has been approved for use by the FDA in correction surgery to treat bunions.
- *Prostheses.* 3D-printed prostheses are a good example of the influence this technique can exert since, on the one hand, they are inexpensive, and on the other, they

are fully customized to the wearer. In addition, they are more comfortable than the traditional prostheses and can be manufactured in a day. Low costs of the 3D-printed limb prostheses are especially important in prosthetics for children since they outgrow the prostheses fast. Moreover, stretchable and expandable 3D prosthetics may soon be available for children that could "grow" with the child. One can find on the Internet "DIY" assistive devices that can be printed by virtually anyone, anywhere. Taking into account the high cost of the traditional prostheses, this leads to a revolution disrupting the prosthesis market.

- *Implants.* Today several manufacturers produce high-quality replacements or implants for spine, hip, pelvis, trachea, 75 percent of a man's skull, and other body parts. The implants are patient-specific as concerns their size and shape determined on the basis of medical imaging data—such as X-ray for bones, computed tomography (CT), and magnetic resonance imaging (MRI) for bone, soft tissue, and blood vessels. The customization of the 3D-printed implants to the patient represents a true personalization. Scientists from the University of Michigan printed airway splints for babies with tracheobronchomalacia that makes the tiny airways around the lungs prone to collapsing. The airway splints are especially significant since they are the first 3D implant made for kids, and they're designed to grow with the patient. Scientists at Princeton University have used 3D-printing tools to create a bionic ear that can hear radio frequencies far beyond the range of normal human capability in a project to explore the feasibility of combining electronics with tissue.

- *Virtual surgical planning.* Imaging techniques are important in medical practice. Introducing 3DP brings an essential improvement in surgical planning. Computed tomography (CT) and MRI images lead to a detailed picture of internal organs and anatomical parts. Their 3D-printed replicas reproduce the size, weight, and texture of organs, allowing surgeons to rehearse complicated procedures on 3D models. The Japanese company Fasotec, bought by Stratasys, developed a biotexture wet model (https://3dprint.com/49992/fasotec-3d-print-wet-model/), realistically mimicking such organs as lungs that allows surgeons and students to practice the operations to be carried out.
- *Three-dimensional printed models in teaching in health care.* 3D Systems (https://3dprint.com/196538/new-3d-printed-organ-models/) is 3D printing precise dental and anatomical models (tumor and organ).
- *Bioprinting.* Manufacturing a human tissue by 3D printing cells is an exciting, booming area of prospective applications of 3DP. The main future aim of 3D bioprinting is to reduce the shortage of supply in the organ donor market. Experts have developed 3D-printed skin for burn victims (http://iopscience.iop.org/article/10.1088/1758-5090/9/1/015006/meta), and companies such as Organovo (https://organovo.com/) are 3D bioprinting with cells to produce tissues, blood vessels, and even small organs. Washington State University modified a 3D printer to bind chemicals to a

ceramic powder, creating intricate ceramic scaffolds that promote the growth of the bone in any shape.

- *Dental.* By combining oral scanning, CAD/CAM design, and 3D printing, dental labs can accurately and rapidly produce crowns, bridges, plaster or stone models, and a range of orthodontic appliances, such as surgical guides and aligners.
- *Pharma.* The possibility to 3D print drugs using downloadable pharmaceutical recipes piped directly into an appropriate 3D printer will have huge implications for the pharmaceutical industry in the same way music downloads have disrupted the music industry. The FDA just approved an epilepsy drug called Spritam (https://www.spritam.com) that is made by 3D printers. It prints out the powdered drug layer by layer to make it dissolve faster than average pills. Imagine how fast the distribution of medication could be with a 3D printer in every second or third pharmacy. Or imagine how different our attitude toward drugs of pharmacies would be if we could print out drugs at homes on our own 3D printers. A new 3D printer Vitae Industries (https://www.vitaeindustries.com) aims to add another piece to this puzzle by giving pharmacies the ability to quickly and easily produce custom doses of drugs for patients based on their specific needs. That means that instead of patients on lots of medication having to remember to take three of one pill, two of another, and so on, they could instead receive their required doses in vastly simplified form.

Genomics and Big Data

By understanding the information processes underlying life, we are starting to learn to reprogram our biology to achieve the virtual elimination of disease, dramatic expansion of human potential, and radical life extension.

—*Ray Kurzweil, The Singularity Is Near*

The cost of genome sequencing has plummeted a hundred-thousand-fold, from $100 million per genome in 2001 to $1,000 per genome today, outpacing Moore's law by three times. Very soon the price point will approach zero. The primary goal that drives these efforts is to understand the genetic basis of heritable traits and especially to understand how genes work in order to prevent or cure diseases. The amount of data being produced by sequencing, mapping, and analyzing genomes propels genomics into the realm of big data. Personal genomics—understanding each individual's genome—is a necessary foundation for predictive medicine, which draws on a patient's genetic data to determine the most appropriate treatments. Medicine should accommodate people of different shapes and sizes. By combining sequenced genomic data with other medical data, physicians and researchers can get a better picture of disease in an individual. The vision is that treatments will reflect an individual's illness and not be a "one treatment fits all" as is too often true today. With large data sets, we will also be able to unlock the secrets of our biology. We'll find insights into and cures for cancer, heart disease, and neurodegenerative disease, and ultimately extend the human life span.

CRISPR or Gene Editing

Every decade or so, there are only a few new ideas that show the potentials of revolutionizing the whole industry of health care and pharma. The amazing genome-editing method CRISPR (https://www.broadinstitute.org/what-broad/areas-focus/project-spotlight/questions-and-answers-about-crispr) has this potential. Researchers have already used gene-editing to create mosquitoes that are almost entirely resistant to the parasite that causes malaria. Some scientists also believe that we will have the chance to edit our cells in our immune systems with CRISPR to improve them against cancer cells and to help them kill these malevolent entities in time. Finally, in August 2017, the Food and Drug Administration (FDA) approved the first-ever treatment that uses gene editing to transform a patient's own cells into a "living drug." Kymriah (https://www.us.kymriah.com/acute-lymphoblastic-leukemia-children/), a one-time treatment made by Novartis, was approved to treat B-cell acute lymphoblastic leukemia—an aggressive form of leukemia that the FDA calls "devastating and deadly." The FDA is currently considering over 550 additional experimental gene therapies. What happens to our healthy human life span as these life-saving treatments demonetize and become universally accessible?

Regenerative Medicine

Perhaps the most significant trends and breakthroughs in the next decade will be in regenerative medicine. Developments have been made growing tissue and even organs in labs to help restore normal functionality in patients. Many of these regenerative therapies take advantage of the advancements made in stem cell research. There have already been breakthroughs that could potentially give us the ability to repair nerve damage or even grow entire organs and limbs.

We are now in the earliest stages of stem cell therapy development. Future therapies will be transformative and, frankly, mind-boggling. Stem cell therapy promises tissue regeneration and renewal and thus a "cure" for everything from blindness to spinal cord injuries, type 1 diabetes, Parkinson's disease, Alzheimer's disease, heart disease, stroke, burns, cancer, and osteoarthritis.

In 2012, researchers at Cedars-Sinai reported one of the first cases of successful therapeutic stem cell treatment. They used patients' own stem cells to regenerate heart tissue and undo damage from a heart attack.

UC San Francisco researchers have safely transplanted a woman's stem cells into her growing fetus, leading to the live birth of an infant with a normally fatal fetal condition. The infant, who had been critically ill during the second-trimester of pregnancy due to alpha thalassemia, is the first patient enrolled in the world's first clinical trial using blood stem cells transplanted prior to birth.

The infant was born at UCSF Medical Center at Mission Bay in February, four months after undergoing the transplant to treat the blood disorder, which is caused by a gene carried by nearly 5 percent of the world's population. Normally, women whose fetuses are diagnosed with alpha thalassemia are given a grim prognosis and often terminate the pregnancy due to the low likelihood of a successful birth. However her birth suggests that fetal therapy, including fetal transfusions, is a viable option to offer to families with this diagnosis.

Augmented Reality (AR) and Virtual Reality (VR)

Augmented reality (AR) is a live direct or indirect view of a physical, real-world environment whose elements are "augmented" by computer-generated or extracted real-world sensory input—such as sound, video, graphics, haptics, or GPS data—whereas virtual reality replaces the real world with a simulated one. While AR lets users see the real world and projects digital information onto the existing environment, VR shuts out everything else completely and provides an entire simulation. These distinctive features enable AR and VR to become a driving force in the future of medicine.

Applications of AR and VR in health care continue to grow and span the spectrum of patient care optimization to improving management and care delivery. Some of the leading examples of AR in health include:

- *Saving lives through showing defibrillators nearby.* The Layar reality browser combined with AED4EU app (https://www.layar.com/layers/sanderl) help locate automated external defibrillators (AEDs) and potentially other first-aid equipment in time of emergency.
- *Helping patients describe symptoms better.* EyeDecide (https://www.eyesdecide.com/) uses the camera display for simulating the impact of specific conditions on a person's vision. For instance, the app can demonstrate the impact of cataract or AMD and thus help patients understand their actual medical state (https://orcahealth.com/).
- *Helping nurses find veins easier.* AccuVein (https://www.accuvein.com/home/) uses augmented reality by using a handheld scanner that projects over skin and shows nurses and doctors where veins are in the patients' bodies. It's been used on more than ten million patients, making finding a vein on the first stick three and a half times more likely. Such technologies could assist health-care professionals and extend their skills.
- *Pharma companies providing more innovative drug information.* With the help of AR, patients can see how the drug works in 3D in front of their eyes instead of just reading long descriptions on the bottle.
- *Augmented reality assisting surgeons in the OR.* Medsights Tech (https://www.youtube.com/watch?v=nWC2Zsh6gZQ) uses augmented reality to create accurate three-dimensional reconstructions of tumors. The complex image-reconstructing technology basically empowers surgeons

with X-ray views, without any radiation exposure in real time. Similarly, HoloAnatomy (https://www.youtube.com/watch?v=5PbAdHAgWA8) and EchoPixel (http://www.echopixeltech.com/) use a wide variety of current medical image datasets, enable physicians to envision key clinical features, and help in complex surgical planning, medical education, or diagnostics.

- *Teaching life skills to those on the autism spectrum.* Brain Power (http://www.brain-power.com/) transforms wearables into neuroassistive devices for the educational challenges of autism.
- *Assisting the visually impaired.* Aira (https://aira.io/) uses deep learning algorithms for describing the environment to the user, reads out text, recognizes faces, or notifies about obstacles. Using a pair of smart glasses or a phone camera, the system allows an Aira "agent" to see what the blind person sees in real time and then talk them through whatever situation they're in.
- *Improving productivity.* Augmedix (https://www.augmedix.com/) provides a technology-enabled documentation service for doctors and health systems so physicians do not have to check their computers during patient visits while medical notes are still generated in real time. Atheer (https://atheerair.com/), a pioneer of the augmented interactive reality (AiR) smart glasses, enables users to view critical work information right in their field-of-view and interact with it using familiar gestures, voice commands, and motion tracking. Users can collaborate with remote

experts via video calls and receive guidance through real-time image annotations to increase efficiency, all while keeping the focus on the task at hand.

Similarly, although the field of VR in health care is relatively nascent, there are examples of VR having a positive effect on patients' lives and physicians' work.

- *Medical training.* Companies such as Medical Realities (https://www.medicalrealities.com/) could elevate the teaching and learning experience in medicine to a whole new level. Today only a few students can peek over the shoulder of the surgeon during an operation, and it is challenging to learn the tricks of the trade like that. With a virtual-reality camera, surgeons can stream operations globally and allow medical students to actually be there in the OR using their VR goggles. ImmersiveTouch (https://www.immersivetouch.com/) offers a VR-imaging platform that allows surgeons to see, feel, and experience minimally invasive surgical pathways to improve surgical precision and patient outcomes. Osso VR (http://ossovr.com/) is working on utilizing the power of VR in the field of orthopedic surgeries, at the same time trying to improve the method of surgical training.
- *Stress and pain relief.* Brennan Spiegel and his team at the Cedars-Sinai hospital in Los Angeles introduced VR worlds to their patients to help them release stress and reduce pain. With the special goggles, they could escape

the four walls of the hospital and visit amazing landscapes in Iceland, participate in the work of an art studio, or swim together with whales in the deep blue ocean. By reducing stress and pain, the length of the patient's stay in the ward or the amount of resources utilized can both be decreased using FirstHand Technology (https://firsthand.com/).

- *Speed-up recovery.* MindMotionPro (https://www.mindmaze.com/mindmotion/) allows stroke patients to "practice" how to lift their arms or move their fingers with the help of virtual reality. Although they might not carry out the actual movement, the app enhances attention, motivation, and engagement with visual and auditory feedback. The app makes the practice of repetitive movements fun for patients. The resulting mental effort helps their traumatized nervous systems to recover much faster than lying helplessly in bed.
- *Improved hospital experience.* VisitU (http://visitu.nl/) makes live contact possible with a 360-degree camera at the patient's home, school, or special occasions, such as a birthday celebration or a football game. Though hospitalized, young patients can relax and still enjoy their lives.
- *Phobia treatment.* Psious (https://www.psious.com/) offers unique treatment for psychological conditions, such as fear of flying, needles, various animals, public speaking, general anxiety, or agoraphobia. With the help of VR, patients get into situations that are fearful for them under the constant control of a physician. Their task is to face their fears and

> gradually let them go while their imagination is helped by VR. Virtually Better (http://www.virtuallybetter.com/) offers among others an exposure therapy for people suffering from anxiety disorders, specific phobias, or PTSD.

In July 2017, the University of Minnesota doctors used VR to prepare for a challenging nonroutine surgery, separating a pair of twins conjoined at the heart. Not only was the life-saving surgery a success, but also the VR prep gave doctors unforeseen insights that prompted them to accelerate the surgery by several months. It won't be long until we refuse to have surgery completed by any human who hasn't prepared in virtual reality using a personalized 3D model.

CHAPTER 3

Patients to Prosumers

It's the summer of 2015, and after completing a keynote address at a health innovation conference in Destin, Florida, I was enjoying a birthday dinner with my family at the Oceans restaurant. The toasts were over, the band was playing a Sinatra classic, and as I leaned over to take the second bite of my medium-rare fillet, I got a sudden sharp pain in my chest. I got the fright of my life; after all, I am a not so fully compliant, pre–diabetic, borderline hypertensive with a cholesterol problem . . . having a steak at a seafood restaurant, and I was 270 miles away from my cardiologist.

I reached for my iPhone that had attached to it the Kardia Mobile EKG. Within thirty seconds, I ran an EKG and got an instant analysis. Normal! Relieved but still anxious, I hit save and sent it to my physician. I got a reassuring call back: "take a Tums and let's review when you get back."

I just saved thousands of dollars on ER charges and freed up the ER to focus on real emergencies, not to mention the instant relief for my family and me. That evening, my smartphone made me a health-care prosumer.

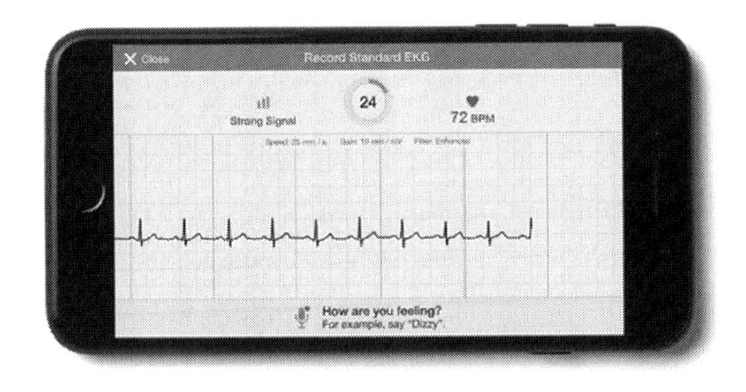

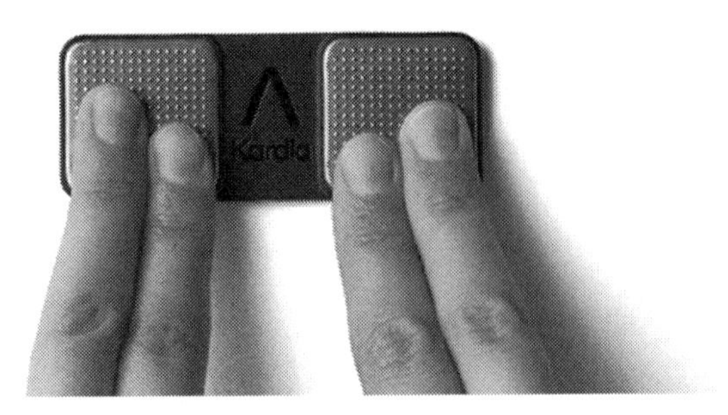

Kardia Mobile EKG

One of the root causes of the ineffectiveness and inefficiencies that we face in health care is the problem of market failure, implying that the fair rules of market principles do not apply and patients often become losers. One of the key causes of this market failure is the information asymmetry between doctors and patients and between patients and insurance companies. This results in a supplier-driven demand situation (that incidentally is unique to health care), the result of which equates to monopolistic or oligopolistic behavior of health-care practitioners. For two millennia since Hippocrates, only physicians were able to acquire and access medical data and make medical decisions. This model of medicine trains physicians to think they know best what's good for the patient. Patients are subjects of health care, not partners. Historically, patients' lack of medical knowledge have placed them at the mercy of practitioners—imagine being instructed when to purchase the next service or product and from whom—making them passive recipients of care as opposed to being active participants, in effect living up to the Latin derivation of patient "patior," meaning "to suffer, bear a burden."

But this is all changing and fast. We live in incredible times. The digitization of knowledge coupled with the convergence of the Internet, connectivity, and smartphone technology means that knowledge is ever-available on always-connected devices. And this includes medical knowledge. The number of people turning to the Internet to search for a diverse range of health-related subjects continues to grow according to a recent Pew study. Eighty percent of Internet users have searched for a health-related topic online, up from 62 percent of Internet users who said they went online to research health topics less than a decade earlier, so much so that Google has partnered with Mayo Clinic to update its catalog of health symptoms by including more detailed descriptions of specific diseases as well as treatment options and symptoms that might warrant a doctor's visit. The cumulative effect of this is that it gives patients more control over their health and social care needs and reduces their dependence on health-care practitioners. Leveraging off the massive increase in health-related searches, Google also developed a series of visualizations to show how health-related Internet searches map to the actual spread of disease. For example, Google shows that in geographic areas where searches for cancer, heart disease, stroke, and depression are high, so are actual occurrences of those diseases. The data also shows trends over time. Google searches for obesity, for instance, have been steadily on the rise for the past decade.

The Deloitte 2016 survey of US health-care consumers also showed the use of social media for health purposes has also risen over the last few years. Not surprisingly, use among millennials is highest and has increased the most among all the generational cohorts. Purposes include learning more about and sharing personal experiences with a specific illness, injury, or health problem; specific prescription medications or medical devices; specific doctors or hospitals; the health-care system in general; health technologies that can help you diagnose, treat, monitor, or improve your health; or other health- or care-related purposes (Deloitte Center for Health Solutions Survey of US Health Care Consumers, 2013 and 2015).

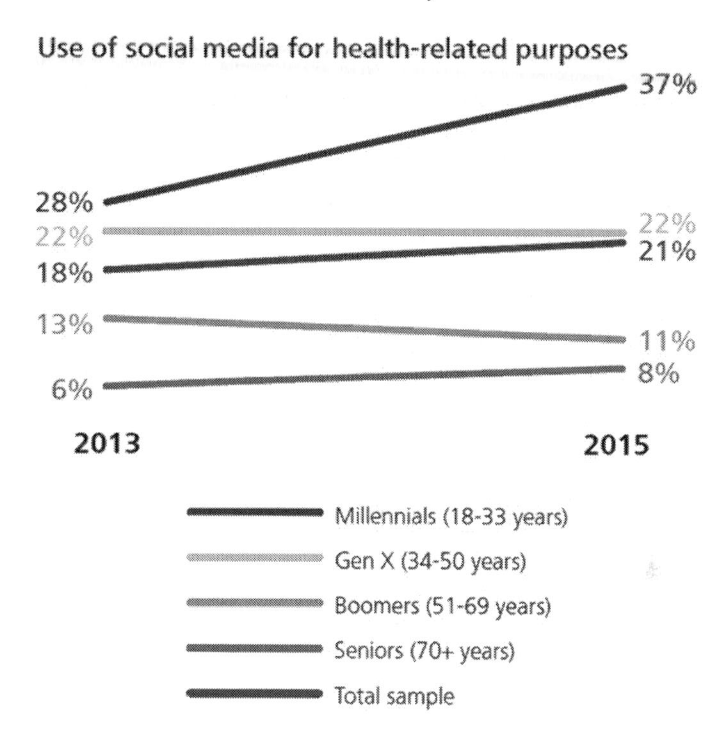

Patients today also have a stronger voice than ever. They are surprisingly well informed through online communities. They share, they learn, and they organize themselves. A community such as PatientsLikeMe groups over six hundred thousand-plus patients who share their symptoms, concerns, experiences with treatment, and healing stories about over two thousand eight hundred conditions. They've generated forty-three million data points about disease, creating one of the largest repositories of patient-reported, cross-condition data available today. They can connect with people with similar symptoms and check what medication was effective and what not. They can get answers to questions they would not dare ask their physicians. They want to talk to patients like them, not only to share a burden but also to try and find a solution that benefits their health.

Similarly, Crohnology.com has a family of close to eleven thousand patients with Crohn's disease from ninety-six different countries, founded out of the realization that data that was gathered outside the doctor's office was just as important as, if not more than, what was learnt inside. Crohnology allows patients to collaborate, share health and treatment information with one another, and track and share their health. Patients learn from one another's experiences in a massively scalable way. This turns the traditional medical model on its head and is probably a better model for medicine for patients living with chronic conditions.

Over and above the instantaneous access to medical knowledge, previously the purview of health-care professionals only, patients are now for the very time also able to compare prices of services. Providers such as the Surgery Center of Oklahoma (https://surgerycenterok.com/) led the way in full price transparency, followed by a number of health plans, provider organizations, state hospital associations, and other groups that have also developed transparency tools.

Furthermore, companies such as XpertDox (https://www.xpertdox.com/) allow patients to compare and contrast the physicians' true expertise for a given disease, symptom, or procedure by evaluating doctors' clinical expertise, research, educational contributions, and leadership roles for each condition. Overall, XpertDox has analyzed billions of data points to provide patients a personalized

360-degree evaluation of a doctor. This enables patients to choose physicians based on objective criteria as opposed to the subjective patient review websites. The Centers for Medicare and Medicaid Services (CMS) also launched Physician Compare to provide useful information about physicians and other health-care professionals who take part in Medicare. This ongoing effort, along with the addition of quality measures on the site, helps Physician Compare provide information to help consumers make informed decisions about their health-care provider choice.

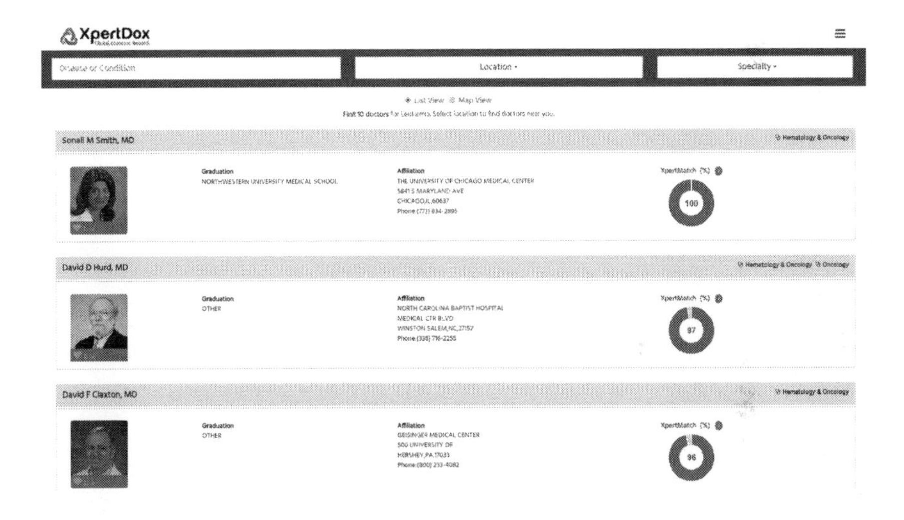

ExpertDox

The democratization of medical knowledge propelled by the advent of the Internet and the rise of social media when coupled with the ability of patients to compare prices and choose physicians based on expertise and efficiency is creating a context where consumerism will thrive and market forces may start to play a role in health care. And although the overall pace of change is gradual, some consumers are making the transition from "passive patients and purchasers" to "informed active health-care consumers." The advent of consumerism is touted as a part of the solution to the quality, cost, and access challenges that we face in the USA and abroad, and there is economic merit to this argument.

However, it is my considered opinion that we have already started to catapult the age of consumerism and have already entered the era of what I call *prosumerism*. Prosumers differ from patients in that they are not passive recipients of health care and they differ from consumers in that they are just not informed purchasers of health care. They actually *produce* their own health care. And the notion of prosumerism or being a prosumer is not a new one. Most of us are already prosumers. If you do your banking online or plan your vacations and purchase air tickets and hotels online or file your own taxes using web-based options, then you are already a prosumer. You are producing your own services and are now competing with your bank teller, travel agent, and accountant. This catapulting to prosumerism in health care is driven by two major forces—one pull and the other a push. But irrespective of what is driving the transition to prosumerism, the consequences of customers transitioning to competitors are dire for health systems and providers.

There are three major forces pushing us toward prosumerism—price (*not cost* as we have no clue what it costs), quality, and access. According to the Commonwealth Fund (see chapter 1), the United States spends far more on health care than other high-income countries, with spending levels that rose continuously over the past three decades. Yet the US population has poorer health than other developed countries. Life expectancy, after improving for several decades, worsened in recent years for some populations, and as the baby boom population ages, more people in the United States and all over the world are living with age-related disabilities and chronic disease, placing pressure on health-care systems to respond. Timely and accessible health care could mitigate many of these challenges, but the US health-care system falls short, failing to reliably deliver indicated services to all who could benefit. In particular, poor access to primary care has contributed to inadequate prevention and management of chronic diseases, delayed diagnoses, incomplete adherence to treatments, wasteful overuse of drugs and technologies, and coordination and safety problems.

Ideally, a higher cost of health care would correspond to a higher quality of care. However, cost is not the only determinant of care. Availability of services also determines accessibility. Currently, the United States has only two and a half practicing physicians for every one thousand people. Because of a rapidly growing population and a much smaller increase in the number of people entering the medical field, the United States faces a shortage of ninety thousand physicians by 2050. Rural America has already been hit the hardest by the physician shortage as a disproportionate number of physicians choose to practice in more urban areas. Efforts to increase the number of practicing physicians cannot quickly yield results enough to meet demand. According to the Association of American Medical Colleges, "Increasing the number of [US] doctors is necessary, but it will not be sufficient. In the coming years, the nation will need to transform the way health care is delivered, financed, and used."

This persistent shortage of skilled workers—from nurses to physicians to lab technicians—will mean hundreds of thousands of positions will remain unfilled. This relative increase in demand is often manifested by increasing waiting times. Patients are waiting an average of twenty-four days to schedule an appointment with a doctor according to a study by Merritt Hawkins of commonly used specialty physicians in fifteen major US cities. The time to schedule an appointment has jumped 30 percent in fifteen US metropolitan areas from eighteen and a half days in 2014 amid a national doctor shortage increased demand. The survey looked at average wait times among five specialties: family medicine, dermatology, obstetrics or gynecology, orthopedic surgery, and cardiology. It will take you forty-five days to see a cardiologist in Boston, one hundred twenty-two days to see a family physician in Albany, New York, seventy-eight days to see a dermatologist in Philadelphia, forty-eight days to see an ob-gyn in Seattle, and nineteen days to see an orthopedic surgeon in Detroit (https://www.merritthawkins.com/uploadedFiles/MerrittHawkins/Content/Pdf/mha2017waittimesurveyPDF.pdf).

The *principle pull toward prosumerism* is the convergence of exponential technologies in medical science, software, hardware, and communications, which have the potential to render the health-care sector unrecognizable in a relatively short amount of time. For the first time in the history of mankind, patients now have unprecedented access to medical knowledge *and* the tools to not only participate in but also direct their own care. These technologies will likely begin to transform how health care is delivered and alter hospital and health system-operating models in the same way that exponentials have already transformed many other industries, changing the way consumers interact with the world. Like other industries, this disruption has the potential to shrink the role and relevance of today's health-care providers and simultaneously help create better, faster, and cheaper services.

According to a report in the *Economist*, in 2017, the number of people without access to banking services fell to 1.7 billion, down from 2.5 billion in 2011, thanks largely to the rise of mobile-payment apps. Physical banks and ATMs can be expensive to set up, especially in rural areas, but the rise of mobile banking has meant that for the first time, hundreds of millions of people living in the developing world now have access to finance. This episode offers a graphic example of how technology can deal with "financial exclusion" by greatly reducing the number of those without access to financial services. Almost inadvertently, the spread of mobile telephony and mobile Internet services has brought hundreds of millions of people into the formal financial system.

Similarly, underpinning the patients' elevation to prosumers is the convergence of smartphone, sensor and AI technologies, notwithstanding the exponential advances in others. An array of sensors (about three hundred currently commercially available and several in development) is now able to give you millions of data points that range from activity levels to emotional status to vital signs to continuous blood labs. The combination of these sensors with smartphone technology and built in artificial intelligence is resulting in a flood of point of care diagnostics—medical testing and monitoring devices that come to you as opposed to you going to a health center.

Why wait seventy-eight days to see a dermatologist in Philadelphia if you're one of the nearly 250 million Americans who own a smartphone and if you have concerns about—or just want to stay on top of—your skincare or skin health? There are several top-rated dermatology apps backed by doctors and scientists.

Download YoDerm (https://yoderm.com/)to get prescriptions for everything from acne to fine lines to longer eyelashes. App users sign up for a consultation or product subscription, fill out a consultation (with photos), and are then connected with a dermatologist who can effectively diagnose them. Users can pick up their prescription or get it mailed directly to their house, and YoDerm follows up with you weekly to see how your treatment is going. Consultation fee is $29.

Spruce (https://www.sprucehealth.com/) lets you take photos of what's bothering you and send them to a licensed dermatologist. It is free to download, and for $40 per assessment (likely $100 less than what you would typically pay to see a doctor), you'll hear back from a doctor within twenty-four hours, and you get a diagnosis, thirty days of follow-up time, and a treatment plan based specifically on your images and prognosis. Spruce doctors can diagnose and treat everything from hair loss to rashes to bug bites and stings. Apps like SkinVision allow you to evaluate and monitor skin conditions at zero cost.

Similarly, why wait 122 days to see a family physician in Albany? For about $25 a consultation, you can use one of several Zipnosis-powered online consulting platforms (https://www.zipnosis.com/) used by some of the USA's top health-care systems. UAB's eMedicine platform is one such product. Their team of clinicians virtually treats more than twenty common medical conditions. Once you create an account and complete a short online interview, their team will review your information and send you a treatment plan generally within an hour.

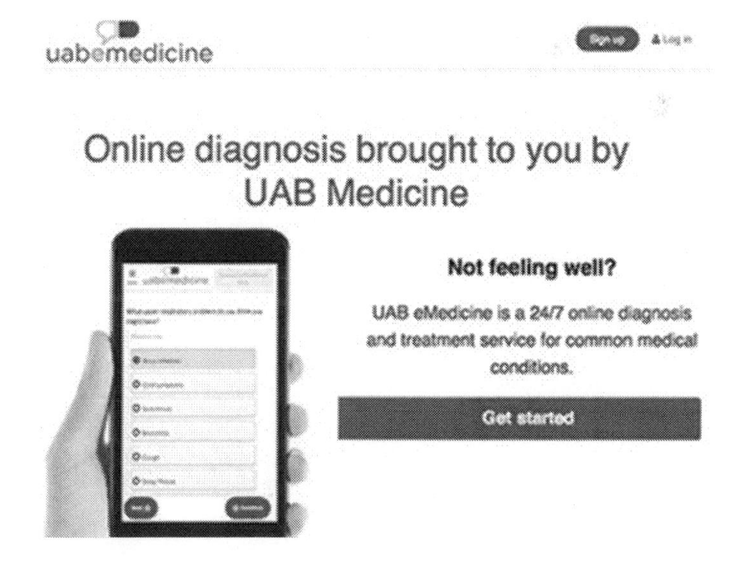

Babylon—coming to the USA soon—combines the ever-growing power of AI with the best medical expertise of humans. It can deliver unparalleled access to health care, including personalized health assessments, treatment advice, and face-to-face appointments with a doctor 24/7. As a result of their partnership with the NHS, patients in the United Kingdom can now see an NHS general practitioner within minutes for free.

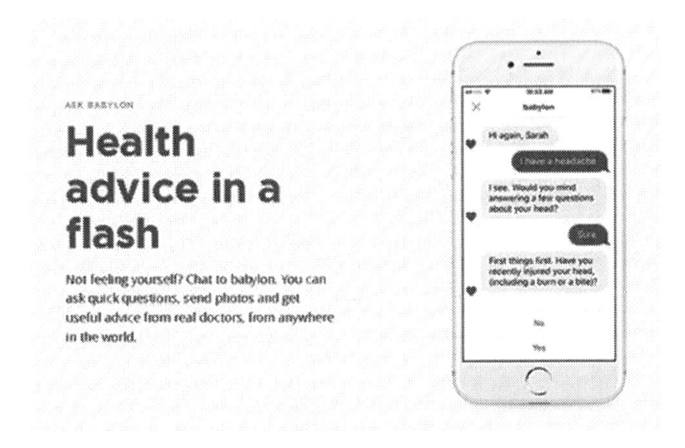

BABYLON

Heart disease is the leading cause of death worldwide, and having to wait forty-five days to see a cardiologist in a city such as Boston does not help. One of the oldest and still the most commonly used diagnostic tools in cardiology is electrocardiography (EKG). Capitalizing on the trend for more mobile health solutions and optimal remote patient monitoring, a number of companies have introduced portable, handheld EKG devices for personal use. WIWE by Sanatmetal is a lightweight, slim device that is roughly the size of a business card and enables you to do a one-lead EKG, pulse rate, and blood oxygen saturation (SpO2). The app provides individuals with a simplified analysis of three major components of an EKG: EKG parameters, detection of arrhythmias (AR), and ventricular repolarization heterogeneity (VH) that is used to determine cardiac muscle health and the risk of sudden cardiac arrest. Each of these categories appears in green, yellow, or red on the screen to indicate the level of deviation detected.

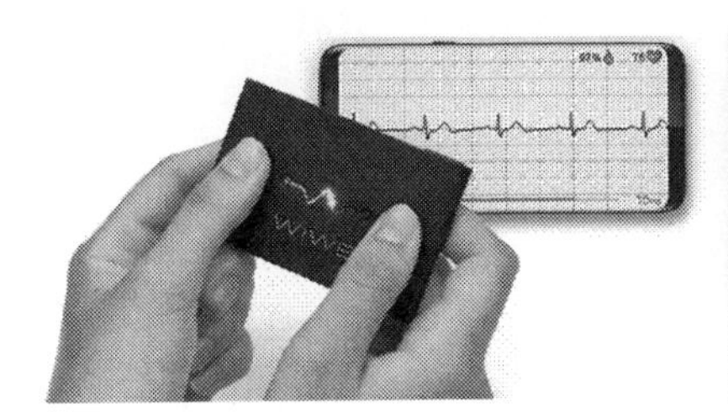

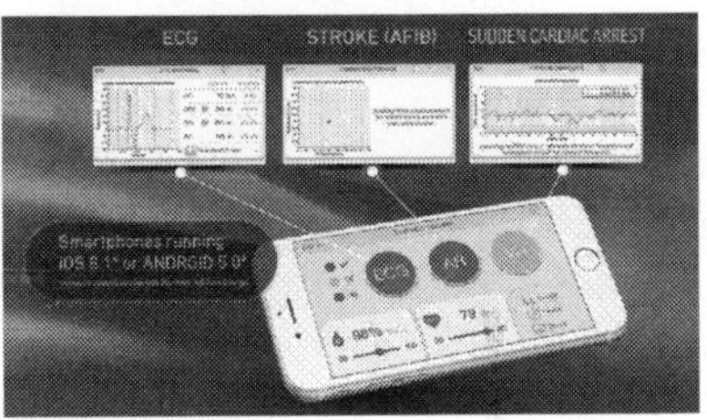

WIWE

Kardia from AliveCor was the first medical-grade EKG technology powered by AI and built for use with the patient's smartphone or Apple watch. Immediate results through FDA-cleared algorithms make it easier than ever to engage patients in their own heart health and provide them with actionable data that matters. The devices allow you to instantly know if your heart rhythm is normal or if atrial fibrillation is detected—the leading predictor of strokes. It also allows you to track your weight and blood pressure and gives you an option to get unlimited history and storage of your EKG recordings, plus customized monthly reports mailed to your home to share with your doctor.

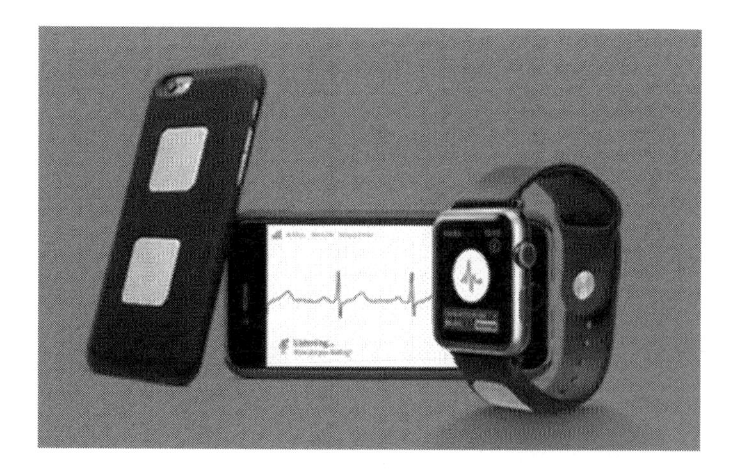

AliveCore's Kardia Devices

The recently released Apple Watch Series 4 also has a ton of additional health features. For starters, it can now detect when you fall by analyzing wrist trajectory and impact acceleration. It will then initiate an emergency call. If it senses you're immobile for 1 minute, it'll automatically call and send a message to your emergency contacts using the SOS feature. While older models of the Apple Watch include an optical heart rate sensor to track calories burned, resting heart rate, and more, the Series 4 offers a few new features. You'll now receive a notification if your heart rate appears to be too low — which could mean that your heart isn't pumping enough blood to the body. In addition, the Apple Watch will now be able to screen your heart rhythm in the background. It'll send a notification to the watch if it detects irregular rhythm, which could point to atrial fibrillation. While the device isn't able to diagnose the issue, it can detect it for you so you can then consult a doctor.

Perhaps one of the most impressive features on the Series 4 is the Food and Drug Administration approved built-in electrical heart sensor. Users can now take an electrocardiogram (EKG) — the first of its kind in a smartwatch. This will measure electrical activity of the heartbeat in order to help diagnose heart disease and other conditions. You're able to take an EKG anytime, anywhere, straight from your wrist by opening the app and placing your finger on the digital crown. Since all of the information is stored in the health app, you'll be able to share the EKG with a doctor, who will be able to see a more detailed picture of what's going on.

These remote-monitoring devices are already proving efficacious as recent research has shown that ambulatory measurements were a stronger predictor of all-cause and cardiovascular mortality than clinic measurements. The study proved that white-coat hypertension was not benign and masked hypertension was associated with a greater risk of death than sustained hypertension.

The idea of a medical tricorder comes from an imaginary device on the science fiction TV show *Star Trek* from the 1960s, which featured fictional character Dr. Leonard McCoy using it to instantly diagnose medical conditions. When Dr. McCoy grabbed his tricorder and scanned a patient, the portable, handheld device immediately listed vital signs, other parameters, and a diagnosis. The good news is that science fiction has become reality. The real life tricorder is here. DxtER (http://www.basilleaftech.com/dxter/) was designed as a device to prove the concept that illnesses can be diagnosed and monitored in the comfort of one's own home by consumers without any medical training. At the heart of DxtER are non–invasive sensors, custom-designed to collect data about a person's vital signs, body chemistry, and biological functions and a sophisticated diagnostic engine based on analysis of actual patient data and years of experience in clinical emergency medicine. It contains algorithms for diagnosing thirty-four health conditions, including diabetes, atrial fibrillation, chronic obstructive pulmonary disease, urinary tract infection, sleep apnea, leukocytosis, pertussis, stroke, tuberculosis, and pneumonia. The device is currently being prepped for FDA clearance and should hit the Lowes store shelves soon for under $200.

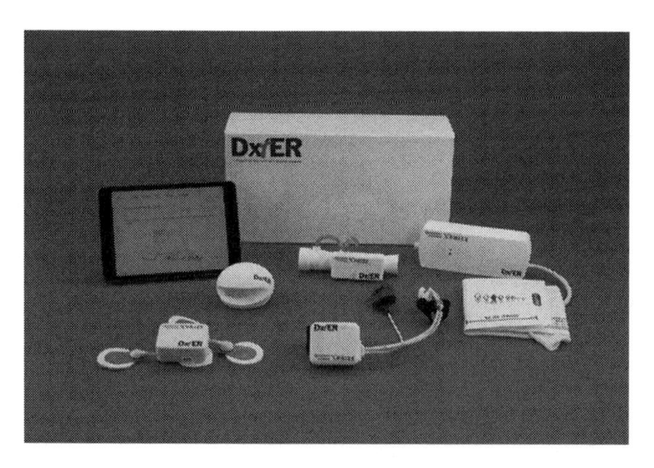

DxtER

Not quite a tricorder but potentially just as disruptive is Tytocare—inspired by a genus of owls known for their smart and excellent hearing, vision, and mobility. Their vision is to transform primary health care by putting health care in the hands of the consumer. Their suite of FDA-approved devices uses guidance technology that ensures anyone can safely and accurately capture exam data. It includes otoscope (ears), stethoscope (heart, lung, and abdomen), basal thermometer, digital camera (skin and throat), and rechargeable battery. The TytoApp allows for live video telehealth exams with your physician or "exam and forward" for later diagnosis. This enables your doctor to remotely diagnose conditions such as ear infections, flu, upper respiratory infections, sinus infections, pink eye, rashes, bug bites, wounds, sore throat, and pneumonia as well to monitor chronic conditions without patients leaving home.

TytoHome

Need an eye test? No need for an appointment. The EyeQue Personal Vision Tracker is the first in-home vision testing solution to combine a cloud-based technology platform, a smartphone application, and a miniature optical scope to form a low-cost, uniquely precise option for people to gather corrective vision measurements whenever and wherever they choose that can be used to order corrective eyeglasses. So now for a smartphone and less than $30, it is possible to become the family or indeed the neighborhood optometrist.

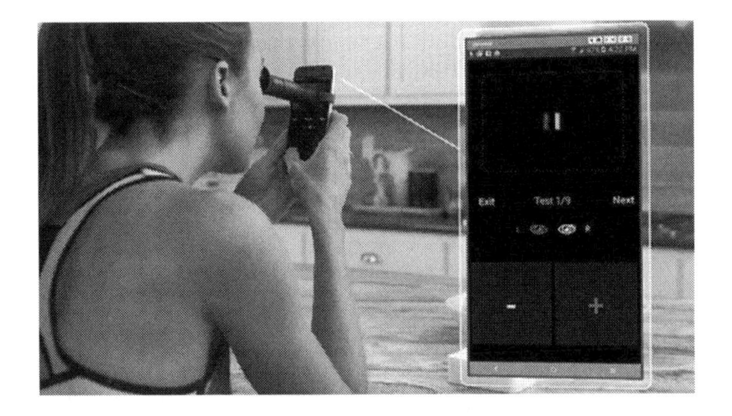

EyeQue Personal Vision Tracker

More recently, there has been a surge in researchers and companies developing sophisticated technology to turn the smartphone into a portable lab, a particularly useful tool in remote and underserved locations. With the number of cellphone users already having passed 7.4 billion, 70 percent of which are in developing countries where there is a critical need for point-of-care diagnostics, this trend is timely. Several cellphone-based devices and smart applications have been commercialized for the monitoring and management of basic health parameters, such as blood glucose, but the possibilities are numerous. A smartphone could be used to test for and begin treatment of diseases, such as hepatitis, malaria, HIV/AIDS, even Zika and Ebola in remote locations. It could help consumers determine if they or their children have the flu before heading out to work or school. Through an mHealth collection platform, it could help clinicians identify outbreaks or treat patients who can't get to the clinic or hospital.

Researchers at the University of Illinois at Urbana-Champaign have developed technology that enables a smartphone to perform lab-grade medical diagnostic tests that typically require expensive large instruments. Costing only $550, the spectral transmission-reflectance-intensity (TRI) analyzer attaches to a smartphone and analyzes patient blood, urine, or saliva samples as reliably as clinic-based instruments that cost thousands of dollars.

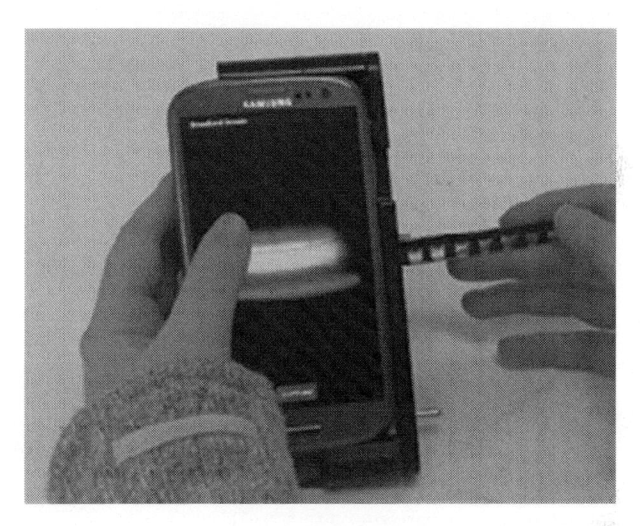

TRI Analyzer

I-calQ (https://i-calq.com/) also uses an inexpensive smartphone attachment to quantify, interpret, and record point-of-care diagnostic testing, creating, in essence, a low cost, portable, easy-to-use system that can replace the traditional medical laboratory. In 2014, i-calQ launched a thyroid disease management product that uses TSH (thyroid stimulating hormone) measurement as well as a stress management system based upon the measurement of salivary cortisol.

Thailand and India have introduced smartphone-based screening of all newborns using i-calQ's technology. Similar programs are running across South and East Asia. In conjunction with local partners and governmental agencies, i-calQ's goal is to test all newborns for congenital hypothyroidism, a serious medical condition that—if untreated—causes permanent severe mental retardation, stunted growth, deafness, and a number of other significant medical problems.

I CalQ's Mobile Diagnostic and Disease Management Solution

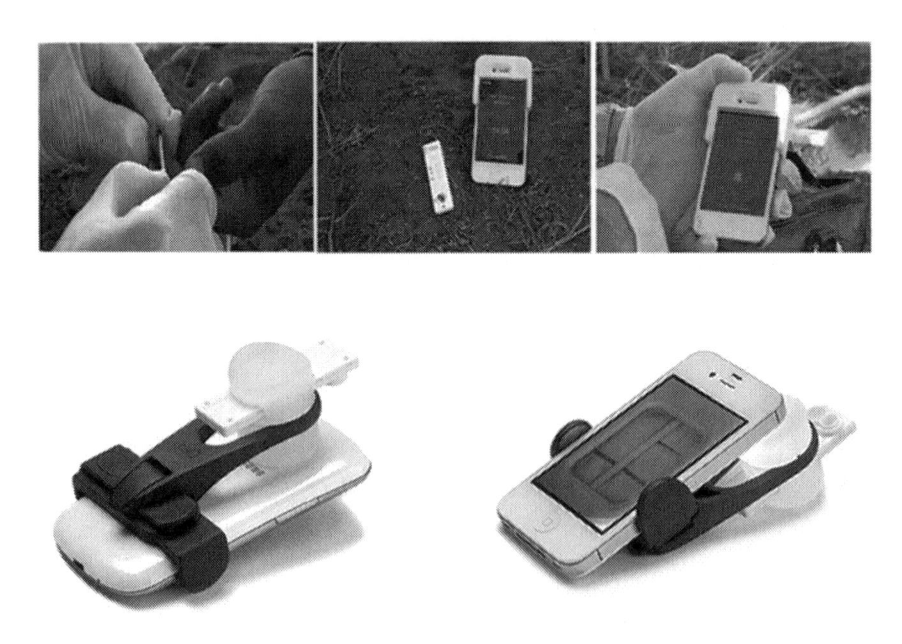

Healthy.io (https://healthy.io/) uses color recognition, computer vision, and AI—in conjunction with a urine sample kit and testing strips—to allow what could be termed a "medical selfie." The user opens the kit, fills the cup, dips the stick, and places it on our patented color board. After waiting for sixty seconds (timed within the app) both the color board and dipstick are scanned, similar to how a QR code is scanned. The image is normalized, and data points are sent to a cloud where they are then classified into the correct clinical result. It currently tests for ketones, leukocytes, nitrites, glucose, protein, blood, specific gravity, bilirubin, urobilinogen, and pH. These indicators span a wide range of pathologies—from urinary tract infection to ketosis, kidney disease, health in pregnancy, and bladder cancer. They will soon be bringing an albumin-creatinine ratio test to the market, which is critical for the seventy-six million Americans (people with diabetes, hypertension, etc.) that need to get their urine tested for signs of chronic kidney disease every year.

In another example of a more consumer-friendly concept, the Australian company ResApp (https://www.resapphealth.com.au/) is testing an app that analyzes a user's cough. The SMARTCOUGH-C study is currently on hold to iron out some glitches but is slated to begin anew soon. This company has pioneered algorithms that accurately characterize the state of patients' respiratory tracts. The team has created a powerful platform for respiratory disease diagnosis and management, which only requires the sound of the patient's cough or breathing and does not require physical contact. With the high-quality microphones in today's smartphones, the platform can be delivered without the need for additional hardware.

The advances in sensor technology coupled with the wealth of data that comes with it is also seeing the introduction of data-driven diagnostics as opposed to the use of biological matter. The traditional home pregnancy test could soon be obsolete, with the introduction of a new smart-watch system that alerts women when they conceive. The team behind the Ava bracelet (https://www.avawomen.com/) described the subtle changes in skin temperature, breathing rate, and pulse and heart rate variability that occur when a woman conceives. This certainly beats waking up every morning and trying to pee on a stick to detect the rise in HCG.

Similarly, Empatica's Embrace bracelet (https://www.empatica.com/embrace/) continuously records physiological signals from multiple sensors and transmits data to a paired smartphone and from the smartphone to Empatica's secure servers via Internet connection. Their advanced machine learning technology helps to predict and identify convulsive seizures and send alerts to caregivers, which is life changing for those affected and afflicted by this condition.

Smartphones are undoubtedly the twenty-first century's doctor bag. They're poised to revolutionize health care the way they have transformed how we listen to music, chat with friends, read the news, pay our bills, and more. The day is not far off when these essentially smartphone-based point-of-care diagnostics would be used by most people, which would empower them to monitor and manage their own health . . . the prosumerisation of health care. The advent of ubiquitous care at costs approaching zero is now a reality.

Converging exponential technologies are also set resolve some of the key challenges that we face in specific disease groups. Let's take the case of diabetes. About 10 percent of the world's population has diabetes, and the number is growing. The American Diabetes Association estimates that the total costs of diagnosed diabetes have risen to $327 billion in 2017 from $245 billion in 2012 when the cost was last examined. This figure represents a 26 percent increase over a five-year period—this for a condition for which we have a pretty good understanding of causation and treatment. Noncompliance and nonadherence to diet, monitoring and treatment increases the complication rate that drives the costs of diabetes management. But technology is about to change this equation as it has the potential to eliminate the disease completely.

Google has partnered with Novartis to launch the smart contact lens to monitor blood glucose from tears and the UCSD has just completed testing a digital tattoo to do the same. These devices will enable patients to painlessly monitor glucose levels in real time using sensor technology and their smartphones.

Smartphones will continuously transmit readings to a bionic pancreas (FDA approved by Medtronic), which will automatically control blood glucose levels by emitting the two hormones that control glucose levels—insulin and glucagon. And if you really cannot resist, there's a food scan available (see Tellspec at http://tellspec.com/) to monitor and control your carbohydrate consumption. Uncontrolled diabetes and the more dangerous hypoglycemia will soon be history, and physicians will be optional.

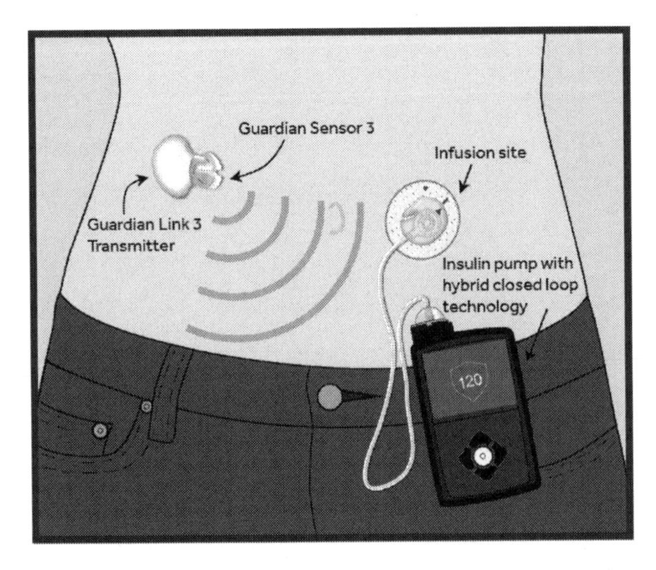

Medtronic Artificial Pancreas

Recently, scientists from the Salk Institute used the CRISPR gene-editing tool to target genes that promoted the growth of insulin-producing cells in mice, and after treatment, the mice were found to have lower blood glucose levels. At the same time, a San Francisco-based company Encellin hopes to change the treatment for type 1 diabetes by using stem cell technology to replace pancreatic islet cells that are the primary defect in type 1 diabetes. The company's investigational device resembles a patch, consisting of donor human pancreatic islet cells residing within two sealed layers of a thin nanoporous material. This material is designed to allow the donor cells to detect glucose levels and secrete insulin as needed, just as normal islet cells would. Both these technologies have the potential to eliminate diabetes at the level of the gene or, at the very least, make you completely insulin independent.

We have seen the similar disruption in the management of peptic ulcer disease that affects about 10 percent of the population. If you were unlucky enough to have this condition prior to the nineties, you most likely would have ended up with surgery with a bill exceeding $5,000. Fortunately, in 1982, Australian scientists identified a bacteria, helicobacteria pylori, as the main cause of ulcers and treatments from midnineties comprised gastroscopies and biopsies to identify the bacteria and a subsequent two-week dose of antibiotics. Total cost is about $2,000. In the 2000s, you could have got a simple blood test to identify the bacteria, and the total cost, including physician lab and medication, is approximately $1,000. Fast forward to today and you can do a simple breathe test with an OTC product (fifteen for $30) or use a smartphone-based diagnostic device and self-medicate with prepacked triple regimen. Total cost is less than $100—physician optional.

This ability of patients to diagnose and manage an illness that just over twenty years ago required a surgeon is an example of what Clayton Christensen described as disruptive innovation, a transformational force that is set to make quality health care both affordable and accessible.

Disruption among providers treating peptic ulcer disease

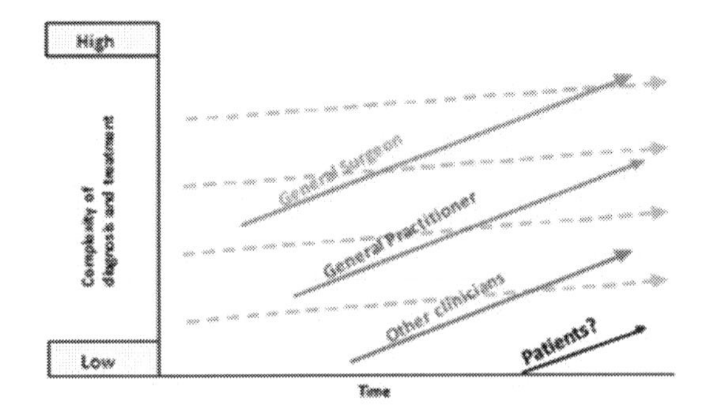

In the coming years, you will have the technology that will enable you to monitor almost every organ system, no matter how difficult to access and no matter what illness you have. We have now entered the era of health care 3.0, driven by technologies that are going to increasingly commoditize medical knowledge and practice and give patients an unprecedented ability to control their destinies—from being a passive patient to an informed, empowered, and liberated consumer; from having your illnesses managed to becoming the CEO of your own health.

CHAPTER 4

Pipelines to Platforms

Distress signals are starting to sound in two of the country's major sectors—retail and health care. Just recently, the discount shoe retailer Payless Shoes filed for chapter 11 bankruptcy and announced it will close four hundred brick-and-mortar stores throughout the United States and Puerto Rico. The announcement comes on the heels of a seemingly unending parade of bad news from traditional retailers in recent months. So far, Walmart, Macy's, J. C. Penney, among others, have all announced significant store closures. Ralph Lauren is shuttering its flagship Polo store, a foot-traffic magnet on Fifth Avenue in Manhattan, the latest step in a massive cost-cutting effort. Big-box office supplies stalwart Staples is reportedly considering putting itself up for sale. Most recently, General Electric, the last original member of the Dow Jones industrial average, was dropped from the blue-chip index and replaced by the Walgreens Boots Alliance drugstore chain. Over the last year, G.E.'s shares have fallen 55 percent, compared with a 15 percent gain for the Dow. For these legacy companies, the stakes continue to rise. From 1965 to 2016, the "topple rate" at which they lose their leadership positions increased by almost 40 percent. Despite a soaring stock market, firm performance as measured by the rates of return on assets of US firms is still one-quarter of what it was in 1965.

So although we live in a time of unprecedented new technological possibilities and worker productivity continues to increase, firms find it difficult to take advantage of these possibilities. While the technological foundations of our world are changing exponentially, firms' ability to adapt is still following a linear path.

Analysts are pretty much unanimous in their assessment of what's ailing traditional firms. Based on McKinsey's research of more than 2,400 publicly traded companies around the world, they estimate that the economic profit generated by technology companies grew a hundred-fold or by $200 billion from 2000 to 2014. There's little doubt that e-commerce companies have delivered enormous gains in efficiency and productivity and have dramatically ramped up the competition, disrupted industries, and forced businesses to clarify their strategies, develop new capabilities, and transform their cultures. The key question then is, as more industries adopt digitally enabled business models—consider, for example, the impact of Amazon in retail, Uber in transportation, and Airbnb in lodging—will this pattern be repeated in other sectors?

Are we starting to see the topple begin in health care? Health systems across the nation are starting to experience some serious pain. MD Anderson Cancer Center lost $266 million on operations in FY 2016 and another $170 million in the first months of FY 2017. Prestigious Partners HealthCare in Boston lost $108 million on operations in FY 2016, its second operating loss in four years. The Cleveland Clinic suffered a 71 percent decline in its operating income in FY 2016.

On the Pacific Coast, Providence St. Joseph Health, the nation's second largest Catholic health system, suffered a $512 million drop in operating income and a $252 million operating loss in FY 2016. Two large chains—Catholic Health Initiatives and Dignity Health—saw comparably steep declines in operating income and announced merger plans. Regional powers such as California's Sutter Health, New York's NorthWell Health, and UnityPoint Health, which operates in Iowa, Illinois, and Wisconsin, reported sharply lower operating earnings in early 2017 despite their dominant positions in their markets.

While some of these financial problems can be traced to capital investments or losses suffered by provider-sponsored health plans, all have a common foundation: increases in operating expenses outpaced growth in revenues. After a modest surge in inpatient admissions from the affordable care act's coverage expansion in the fall of 2014, hospitals have settled into a lengthy period of declining hospital admissions.

For the past several decades, the consensus mitigation strategy among hospital and health-system leaders has been to achieve scale in regional markets via mergers and acquisitions—first in the 1990s, in response to the rapid adoption of HMO-style managed care plans and then again in 2010 in response to the Accountable Care Act (ACA) and presently amid the threat of repealing the ACA. In the case of the advent of HMOs and the repeal of ACA, their theory was that by consolidating their market power, they could make up for shortfalls arising from insurers attempts to decrease utilization. In the case of the ACA, their theory is that they will be better capable of capitalizing on expanding markets and a move toward value-based care. As a result of this only approach to external threats, more than half of the hospital markets in the United States have reached a level of concentration that, in other sectors of the economy, would (and should) provoke an antitrust inquiry or lawsuit.

This response is very typical of most businesses that are based on a pipeline model. Pipeline businesses create value in a linear fashion with centrally employed staff and owned assets. They use a step-by-step arrangement for creating value with producers at one end and customers at the other. Firms create products, push them out, and sell them to customers. Value is produced upstream and consumed downstream. There is a linear flow, much like water flowing through a pipe. We see pipes everywhere. Every consumer good that we use essentially comes to us via a pipe. All manufacturing (including cell phones) runs on a pipe model. Television and radio are pipes spewing out content at us. The hotel industry is a pipe-selling room. Health care is a pipe where providers deliver care to patients.

Typical Pipeline

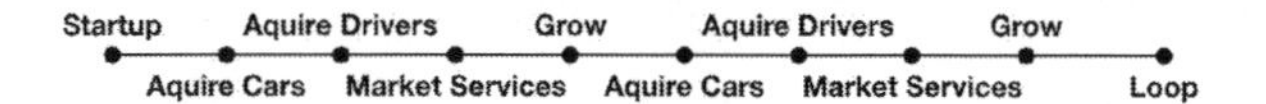

Exponential Growth limited by capital for assets, and need for employees.

Pipeline businesses have traditionally scaled in one of two ways. Some expand by owning and integrating a greater length of the value creation and delivery pipeline, for example, by buying upstream suppliers or downstream distributors—vertical integration. This category represents deals such as CVS Health and Aetna, Optum and DaVita Medical Group, and Cigna and Express Scripts. These couples are trying to lower costs of the supply chain by owning more of it and extending those savings to consumers. Others expand by widening the pipeline to push more value through it—horizontal integration. When hospitals add additional rooms either through building or mergers and acquisitions, this is an example of horizontal integration.

Hospital Pipeline

HOSPITAL

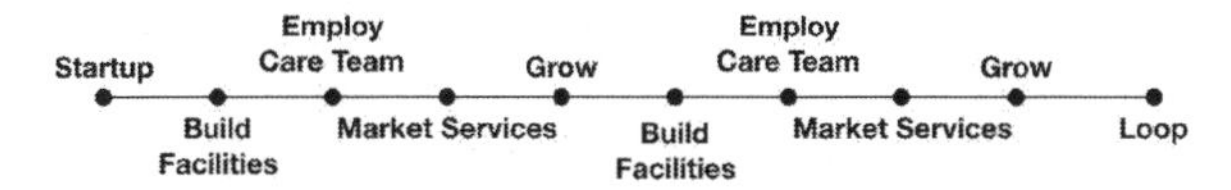

Exponential Growth limited by capital for assets, and need for employees.

Scaling pipelines are resource intensive, while sustaining pipelines in a competitive environment depends on the business's ability to control scarce and valuable—ideally, inimitable—assets. In a pipeline world, those include tangible assets, such as real estate (hospitals or beds), and intangible assets, such as physicians' expertise. This is the logic that drives the mergers and acquisitions agenda in health care and the resulting increase in market concentration, which is tantamount to creating hospital monopolies and oligopolies. The obvious outcome of this is increased prices. James Robinson of the University of California found, on average, procedures cost 44 percent more in hospital markets with an above-average degree of consolidation. On the quality front, an analysis of hospitals in twelve cities by Joe Carlson of Modern Healthcare found, as so many others have, that "there is no consistent relationship between hospitals spending more to perform a procedure and their achieving better patient outcomes." The net effect of the current pipeline-based approach to health care is the nonsustainable cost structure of hospital systems and the rising costs, poor quality, and limited access to care that is so pervasive.

Thankfully, with the rise of mobile and connected technology has come a new approach to business modeling, which profoundly reduces the need to own physical infrastructure and assets and enables quicker, cheaper, and faster scaling. We call these platform businesses, and you don't need to look far to see examples of platform businesses—from Apple to Uber to Airbnb, whose spectacular growth abruptly upended their industries. Uber is the world's largest transportation company and owns no cars, Airbnb is one of the world's largest hospitality companies and owns no rooms and beds, and Alibaba is the world's largest retail company and owns no inventory. Platform businesses have invaded and conquered established legacy companies in a matter of months with none of the resources traditionally deemed essential for survival, let alone market dominance. They achieved this by leveraging the power of platforms.

In 2007, the five major mobile-phone manufacturers—Nokia, Samsung, Motorola, Sony Ericsson, and LG—collectively controlled 90 percent of the industry's global profits. Apple was a weak, nonthreatening player. But by 2015, the iPhone singlehandedly generated 92 percent of global profits, while all but one of the former incumbents made no profit at all.

How can we explain the iPhone's rapid domination of its industry? And how can we explain its competitors' free fall? Nokia and the others had classic strategic advantages that should have protected them: strong product differentiation, trusted brands, leading operating systems, excellent logistics, protective regulation, huge research and development budgets, and massive scale. For the most part, those firms looked stable, profitable, and well entrenched.

As we'll explain, Apple overran the incumbents by exploiting the power of platforms. Platform businesses bring together producers and consumers in high-value exchanges. Their chief assets are information and interactions, which together are also the source of the value they create and their competitive advantage. Apple conceived the iPhone and its operating system as more than a product or a conduit for services. It imagined them as a way to connect participants in two-sided markets—app developers on one side and app users on the other—generating value for both groups. As the number of participants on each side grew, that value increased—a phenomenon called network effects that is central to platform strategy. Apple reported that its app store generated over $26.5 billion in revenue for developers in 2017, which was up about 30 percent year over year. This means that the app store created approximately $11.5 billion in revenue for the company.

The success of Apple and other envied companies such as Uber and Airbnb have made the platform-business model the holy grail of business models. Platforms help to sell products or services, to generate content, and so on. But the platform owner as such does not manufacture the products that get sold (e.g., eBay, Alibaba). They do not provide the services that get offered on their platform (e.g., Uber and Airbnb). They do not create the content that gets generated each day (e.g., Facebook and Twitter). They create value by facilitating asset sharing—in the case of Airbnb, the supply side (homeowners) to increase utilization of their existing assets (unit or room or house) in order to generate additional income and the demand side (travelers) to find cheaper, more individual accommodation and when all hotels are booked out. Airbnb extracts value by charging a percentage of each booking. They don't need to design, build, and own a large amount of their real estate so they can scale quickly and cheaply.

Platform Example

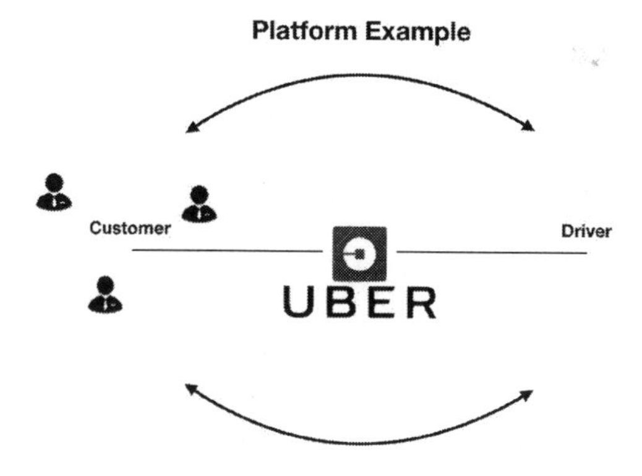

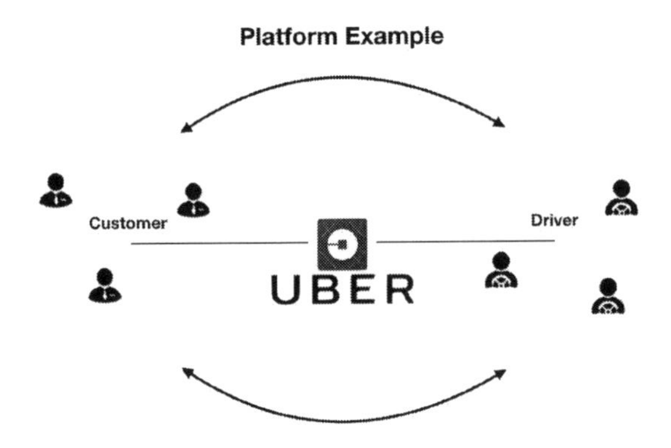
Platform Example
Customer
Driver
UBER

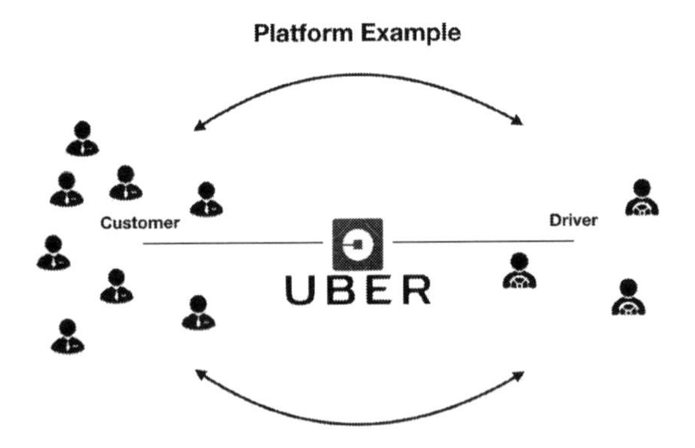
Platform Example
Customer
Driver
UBER

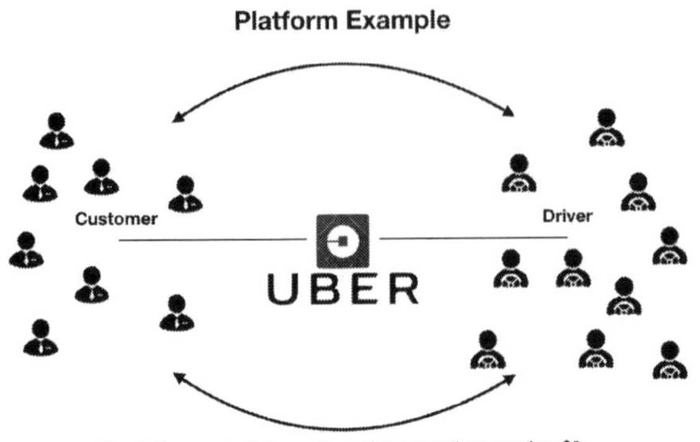
Platform Example
Customer
Driver
UBER
Ready for exponential growth, cost to expand approaches $0.
Growth is directly proportionate to demand.

It is exactly the same model with Uber connecting drivers and passengers. For the platform-business model, the concept of network effects is of utmost importance. Drivers and passengers are on different sides of the platform. Uber would create very little value for a passenger if there were hardly any drivers. Waiting times would be frustratingly long. Equally, for drivers, the platform would have little value without a sufficient number of riders. Idle times for drivers would render the platform of little value. The value of the network increases with the number of cross-site participants—more drivers attract more passengers (lower prices, faster pickups, wider geographical coverage) and more passengers attract more drivers (increased demand, less downtime). This accounts for the explosive growth of platform businesses.

Apple's success in building a platform business within a conventional product firm holds critical lessons for companies across industries. Firms that fail to create platforms and don't learn the new rules of strategy will be unable to compete for long. This is particularly true for health-care organizations that are locked into the traditional pipeline business model. Yet despite health care's remarkable track record holding out against the tides of change, there are finally signs that the power of platforms that have radically transformed some industries is poised to transform health care in three dimensions: administrative automation, networked knowledge, and resource orchestration.

Airbnb's platform makes it ridiculously easy to turn your house into a hotel. They have organized, scaled, and automated the myriad administrative details involved in such a way to minimize the barrier to entry and limit the potential for error, miscommunication, or unwarranted variation. They'll walk you through where to put the square footage and what kind of photos entice the most interest, and they use predictive modeling to show you the best and worst days of the year to get a booking at a particular price. They've even partnered with H&R Block to streamline tax filing.

In health care, the need for administrative automation is viscerally felt, and the potential for alleviating the burden and draining cost from the system is significant. On average, doctors spend just 60 percent of their time each year seeing patients and documenting their care, the other 40 percent processing administrative documents and chasing down missing lab and imaging orders. When providers do what Airbnb hosts have done and set up shop on a network, then much of the work that consumed their day can be automated and dispatched at scale so they can focus on delivering care. A network-based service can, in aggregate, take on administrative tasks such as medical claim submission and posting and get continuously smarter and more efficient with feedback from the network.

With a platform, the critical asset is the community and the resources of its members. In this networked world, the traditional consumer becomes an active producer who adds knowledge and value to the system in a positive feedback loop. By enlisting the driver as a data source, Waze revolutionized the average commute, aggregating data in real-time to flag traffic jams and suggest alternative routes. The app is now so effective that 70 percent of the time, Waze registers a traffic accident before a 911 call is made.

As more patient data gets liberated from isolated systems and added to networks, comparable knowledge and value can be generated for health care. When the American Congress of Obstetricians and Gynecologists (ACOG) expressed concern about the number of women with hypertension who, not yet knowing they're pregnant, continue to use ACE inhibitors that cause serious malformations in fetuses, Athena ran a real-time query of the sixty-three million patient records on their network and identified sixty-two thousand women of childbearing age who were prescribed ACE inhibitors and, therefore, at potential risk. They were able to alert the women's doctors, suggesting they prescribe a different hypertension drug or urge their patients to get on effective contraception. This kind of network medicine can transform care delivery by aggregating knowledge across a vast network and close the gap between that knowledge and appropriate intervention.

Uber brought the taxi industry to its knees by figuring out how to extract new value from excess capacity in the system. They started back in 2009 with black car limos that were essentially the academic medical centers of transportation: extremely expensive and mostly empty. By tapping into that excess capacity and making it available on a network, they generated new market demand, which led to UberX and a thriving community of 160,000 drivers conducting one million rides a day.

Think of the potential for this kind of orchestration of resources in health care where waste accounts for an estimated trillion dollars annually. On any given day in America, 40 percent of hospital beds lie empty (60 percent in rural areas), their enormous fixed costs weighing heavily on the system. Or take medical appointments. In a recent Commonwealth Fund study of patients around the world, 52 percent of Americans said they couldn't get a same- or next-day appointment with their provider when they were sick. When Athena (http://www.athenahealthcare.com/) looked across its network of eighty thousand providers, just 4.4 percent of all appointments were shown as "available" in the next thirty days. But looking back thirty days, just 17.5 percent of all slots were used, suggesting there's a systemic problem in how our industry manages this critical area of patient access. After shadowing fifty primary care physicians across the country, it was observed that 70–80 percent of the PCPs work output is direct waste: inputting computer order entry; prescription processing; composing the billing invoice; filling out mammogram requisitions, including typing in the date and location of the last mammogram; working an inbox full of random notifications and disconnected results; and typing the visit note. So the key question then is "What percentage of the time are health professionals optimally using their skills, that is, working at 'top of license' (TOL)?" My guess is that at best only 20–30 percent of a primary care specialist's day is spent on direct clinical care. My hypothesis is that TOL time (TOLT) varies across specialties being higher in procedural specialties but significantly lower than what it should. The good news is that the opportunities and upside for connecting unsatisfied or latent demand to unused capacity are virtually endless in a platform based model.

Platform Example
Patient
Care Team
TeleHealth

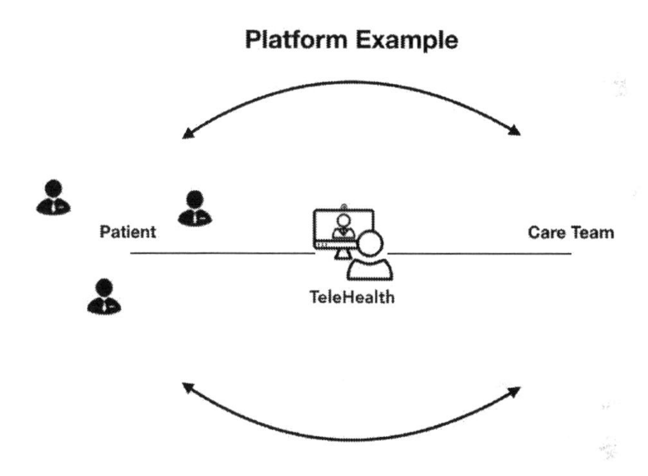
Platform Example
Patient
Care Team
TeleHealth

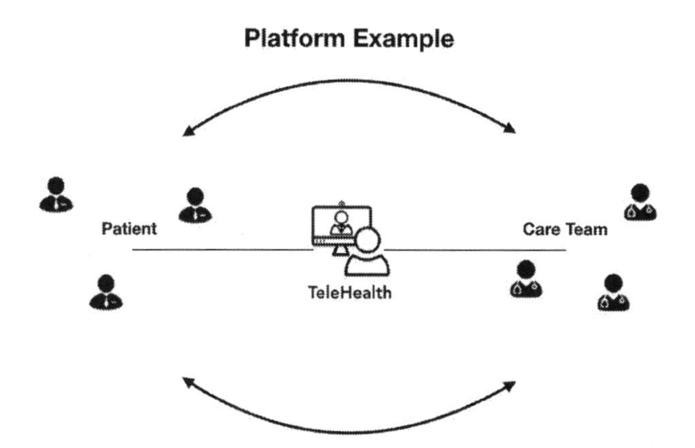
Platform Example
Patient
Care Team
TeleHealth

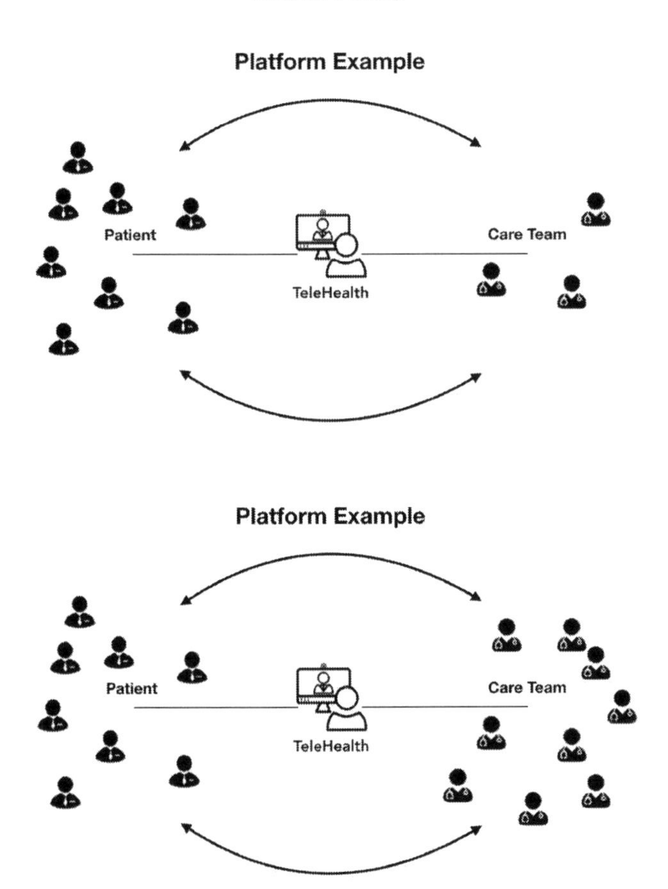

By all accounts, health care in the United States is at a critical juncture, with an urgent need to bend the cost curve, slowing the rate of cost increases while needing to improve access to care. While the government attempts to mandate transformation through maddeningly complex and largely untested models for driving savings and efficiency, we might best be served by looking around us at market-driven models that are transforming the way we shop, travel, meet friends, listen to music, and more. TeleVisits have the ability to be a game-changer in the way we experience medicine. According to the American Medical

Association (AMA), they could replace up to 70 percent of all health-care visits. Almost everyone can see the potential. The tipping point is coming soon, and once it does, the power of the network effect will be profoundly felt by providers and patients alike.

Concomitant with the financial sustainability issues being experienced by hospitals, we are seeing one of the biggest shifts in health-care delivery over the last few decades—the rise of telemedicine. Telemedicine is on the verge of rapid, widespread growth and adoption as the forces that have fueled its rise so far—mainly, the growth of value-based care and patient convenient access—continue to gain momentum. Additional drivers are poised to push telemedicine's expansion beyond what we've seen to date—advances in telemedicine technology, evolution in legislators' and regulators' views of telemedicine, and providers' and insurers' relentless efforts to provide cost-effective care with high-quality results. This is amplified by the growing demand for convenience, innovation, and personalized care among health-care consumers around the world.

The regulatory landscape is beginning to change with the Veterans Administration—the country's largest health-care provider—having recently finalized the work on a rule that overrides the state licensing restrictions. In this way, clinicians can treat veterans anywhere in the country. At the state level, the American Medical Association is advocating for the interstate medical licensure compact. This creates a new pathway to expedite the licensing of physicians already licensed to practice in one state who seek to practice medicine in multiple states. The compact promises to increase access to health care for individuals in underserved or rural areas and allow patients to more easily consult medical experts through the use of telemedicine technologies.

At the time of publication, twenty-two states have adopted compact legislation, and eight more have proposed legislation at this time. When coupled with the soon-to-be ubiquitous availability of wireless 5G and video-capable low-cost devices and increasing demand (50–65 percent of Americans say they would be willing to try a telehealth visit; net promoter score for telehealth visits is in the midseventies, which is incredibly high) propelled by poor access and increases in high deductible plans and health savings accounts, this represents a tipping point in the transition to platform-based care delivery. The global telemedicine market is expected to expand at a compound annual growth rate of 14.3 percent over the next five years, eventually reaching an astounding $86.6 billion by 2020, a yearly increase of 20.8 percent over a five-year period.

So who are health care's Ubers? Well, the first that comes to mind is Uber. They recently launched Uber Health that allows a health-care worker to book a ride on demand or schedule a future ride for a patient. The passenger is alerted by a text or phone call with trip details. Every year, 3.6 million Americans miss doctor appointments because of a lack of reliable transportation. Using "rideshares" is seen as a potential means to drive down health care costs by providing rides to preventative health-care services and using emergency transit less. One study found when Uber enters a city, ambulance use decreases by at least 7 percent. Companies such as Veyo are targeting Medicaid and Medicare patients, and Circulation is offering a platform connecting ride-share services such as Uber and Lyft for nonemergency medical treatment.

The platform-based (or virtual) care delivery space is becoming a dynamic one and is quickly becoming crowded with competitors, and we can expect to see big shifts in the coming years as companies jostle for position and test new business models. Teladoc Inc. (https://www.teladoc.com/) is one of the world's largest telehealth company that uses telephone and video conferencing technology to provide on-demand remote medical care via mobile devices, the Internet, video, and phone. It has over 3,100 licensed health-care professionals on the platform and offers a median response time of ten minutes anytime, anywhere. Revenue growth in 2017 was 89 percent over 2016 and Q1 2018 revenue growth was 109 percent. American Well (https://www.americanwell.com/) offers software, services, and access to clinical services to health plans, employers, and delivery networks. Their mobile and web service connects doctors with patients for live, on-demand video visits over the Internet, and they also handle all the administration, security, and record keeping that modern health care requires. Their recent acquisition of Avizia that provides virtual access to doctors inside of hospitals and health systems across the country will now help patients avoid a doctor's office visit to more tertiary services, offering a menu of services from partner physicians and hospitals around the world.

Doctor on Demand, a national health-care service delivered through telehealth technology, has passed one million video visits. They provide access to board-certified physicians, psychiatrists, and licensed psychologists through video visits as well as integrated lab services for an advanced range of telehealth services. As the company continues to grow, it projects to reach two million visits by summer 2019. Launched in December 2013, Doctor on Demand has seen triple-digit growth year over year.

Then there are companies such as Candescent Health (https://www.candescenthealth.com/) that is virtualizing radiology scans. So a radiologist who specializes in pediatric lungs can spend all his time reading pediatric lungs instead of wasting time with knees and elbows. And a child in rural Wyoming can have her lung X-ray read a thousand miles away by the best pediatric lung radiologist in the country.

The list of platform-based care delivery companies is endless with about 71 percent of health-care systems in the USA already using telehealth or telemedicine tools to connect with patients in the inpatient and ambulatory settings. According to Becker's Hospital Review, telemedicine was set to attract seven million patient users by 2018. The same study predicts that the market will grow by 14.3 percent by 2020. The total market value is predicted to hit $36.2 billion by 2020. In 2018, the number of United States health facilities offering telehealth will be almost twice what it was in 2016.

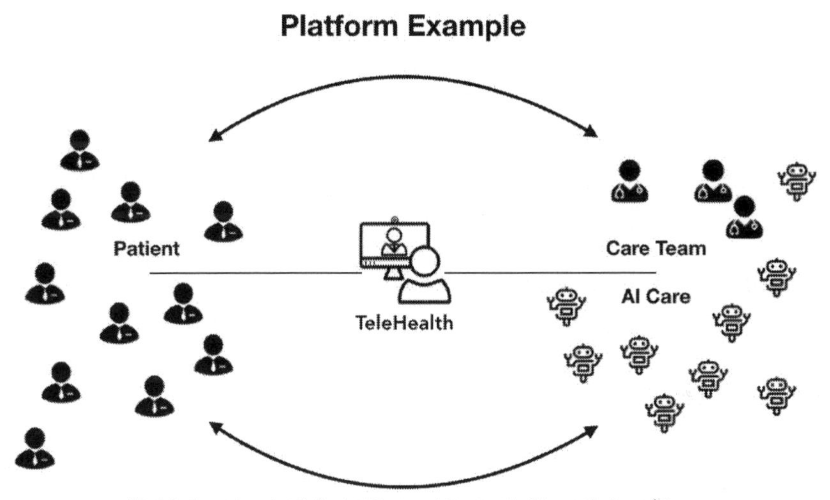

Healthcare is also one of the industries that has the potential to scale significantly faster and at marginal costs approaching zero relatively to traditional platform based companies like Uber and Airbnb. This is because we are uniquely positioned to replace the supply side with AI. Babylon is one such example. Another is Teledoc which has implemented the Oncology Insight with Watson machine learning-powered second opinion service for experts providing cancer care. Utilizing the Watson cognitive computing and machine learning capabilities enhances their doctors' ability to provide the diagnosis and treatment plan for a given patient by providing data-driven suggestions to oncology experts looking for a second opinion. One-third of Watson's rendered opinions come with a recommended change in diagnosis and two-thirds come with changes in treatment. Leveraging Watson in this case improves the quality of care as well as the access to care as productivity increases due to the physician augmentation. It's just a matter of time before advances in deep learning enable us to move from physician augmentation to physician replacement. When that happens it will become possible to scale instantaneously at zero cost thereby transforming traditional care delivery and business models.

The hospital of the future is also set to be platform-based, and the transition has begun. Last year, Mercy Virtual Care opened a first-of-its-kind facility that could redefine patient care. Providing care to patients both nearby and far—but none in the $54 million first-of-its-kind facilities itself—330 specialized medical professionals monitor 2,431 patient beds, of which 458 are occupied by the critically ill. Physicians are seated at a console with six computer monitors filled with a wealth of data to enable them to better assist bedside providers. Secure web cameras allow them not only to see what's going on but also to be seen by those on the other side, whether in one of Mercy's traditional hospitals, a physician's office, or in some cases, the patient's home.

These more efficient digital hospitals are emerging as critical hubs in integrated health-care networks, holding the potential to drive greater efficiency, improve quality of care, and provide access for more people than ever. Whether by newly built or retrofitted existing buildings, digital hospitals promise to boost efficiency and quality through better integration with all sources of care and enable deployment of eHealth systems to provide online information, disease management, remote monitoring, and telemedicine services that can extend the reach of scarce medical resources and expertise. Digital hospitals provide faster and safer throughput of patients, creating more capacity through process efficiencies while containing costs.

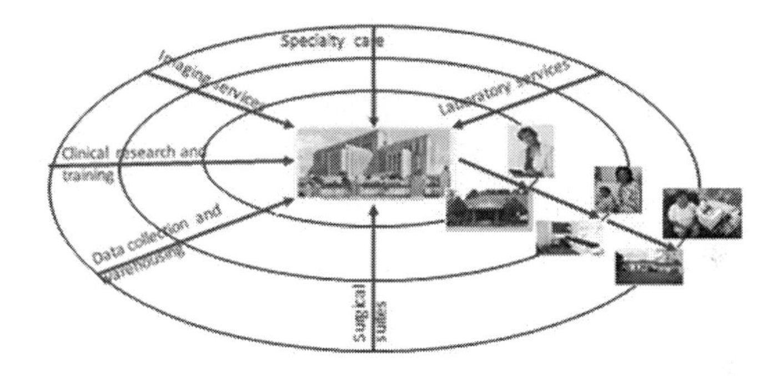

The Decentralization That Follows Centralization Is Only Beginning in Health Care

Hospitals are not just being disrupted by the virtualization of medical knowledge. Exponential converging technologies are also resulting in the dematerialization and commoditization of medical equipment that was previously only available in hospitals. This disruption is mirroring the disruption that we have seen in the computer industry, moving from centralized mainframes in the 1950s to present day where kids of two or three years of age can now play with iPads.

Take for example Tribogenics (http://tribogenics.com/), made the first major breakthrough in X-ray science since it was first invented over one hundred years ago, using triboluminescence-based X-ray technology.

In partnership with Nikon, they are in the process of developing a sub-$10,000 handheld medical X-ray machine. This would have obvious and widespread negative implications for hospital-based radiology departments and positive implications for the developing world where there's little existing medical imaging infrastructure and could offer yet another opportunity for emerging economies to leapfrog more economically advanced nations on the health-care front as handheld and smartphone-related tech enables cheap, high-quality medical imaging and diagnostics.

Ultrasound machines that can easily cost upward of a US$100,000 are also now being commoditized. The immense power of small pocket-sized computers is enabling clever engineers to develop a myriad of portable technological devices that, until now, were expensive and large. One such innovation is called the Butterfly iQ (https://www.butterflynetwork.com/), a small ultrasound device that can display clear black-and-white pictures on an iPhone. Similarly, Clarius (https://www.clarius.com/) has developed a handheld ultrasound scanner that connects wirelessly to a smartphone app, available through iOS and Android app stores, to display the image. When combined with AI, these devices will soon have the potential to transform lay individuals to ultrasonographers, historically highly specialized professionals who are generally hospital and office based.

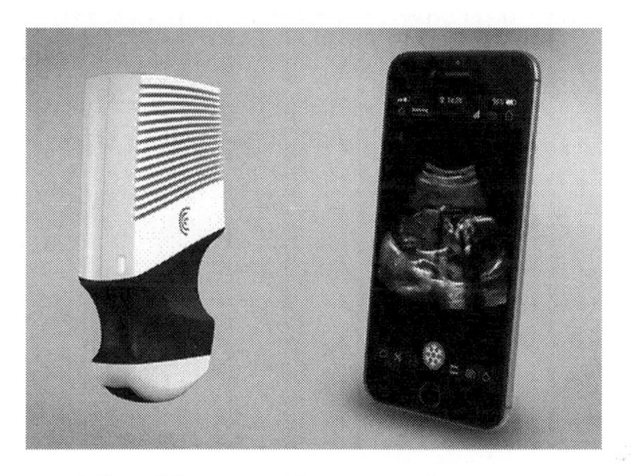

The Clarius Ultrasound Device

Not to be outdone, Philips, one of the leaders in medical ultrasound devices, has also recently launched a small handheld portable scanner, available through a monthly subscription model of $199 per month.

CurveBeam's Cone Beam CT Imaging Equipment

CurveBeam (http://www.curvebeam.com/), a Philadelphia-based company, designs and manufactures Cone Beam CT imaging equipment that is starting to revolutionize the industry with point-of-care CT imaging. The compact systems provide radiology and orthopedic specialists with the three-dimensional bone detail of the orthopedic extremities. They are also the standard of care for advanced orthodontics and oral surgery treatment planning. The systems can be plugged into a standard wall outlet and have minimal shielding requirements; hence, they can be placed in locations convenient to the patient. Radiation dose to the patient is also significantly less than a conventional CT scan. It's just a matter of time before the full suite of CT imaging becomes available in decentralized doctors' offices. It is also conceivable that in the not too distant future, we may be able to access preventive CT imaging in places such as malls, pretty much that same way that Bodyo has commoditized the general medical examination.

Tech firm Bodyo (http://www.bodyo.com/) unveiled their revolutionary AIpod in January 2018. Visitors will be able to step into the futuristic health pod and track their blood pressure, blood sugar, fat mass, muscle mass, hydration, height, and weight for free. The pods are the first of their kind and will be rolled out to gyms, hospitals, pharmacies, malls, and offices in Dubai and across the United Arab Emirates, empowering communities to take charge of their health, prevent serious illnesses, and lower health insurance at a time when obesity has doubled; diabetes is a growing concern globally, and physical activity rate has dropped. Once you have created a profile on the Bodyo app, it takes the data from the Bodyo pod or smart devices and provides individually tailored solutions suited to your body type, lifestyle, and health needs and helps you reach set goals. They include online coaching; predesigned programs, including fitness exercises; remote professional assistance from trainers, nutritionists, and doctors; nutrition and dietary advice; and activity and sleep. The app can connect to more than fifty wearable devices and is compatible with mobile phones, weighing scales, cholesterol meters, accelerometers, GPS pulse oxymeter and heart rate ECG glucometers, etc. It also sends messages and personalized reminders to keep you on track and connects you to like-minded individuals within the Bodyo community, with similar health and wellness goals.

Bodyo AIPod

The stethoscope has long been a revered symbol of medicine, but now the centuries-old device has a new competition with the advent of digital and smartphone-based intelligent stethoscopes.

The Eko DUO is the first FDA-cleared device to combine the digital stethoscope and EKG. The marriage of EKG and digital stethoscope technology into a compact, the handheld device offers unprecedented insight into cardiac function, now enabling patients with heart and respiratory disease to monitor themselves at home.

Eko Duo

Similarly, Steth IO (https://stethio.com/) is a stethoscope for your iPhone. Steth IO is essentially a phone case that is designed to channel the sounds of a patient's heartbeat or breathing into a phone's microphone. The Steth IO app then processes that information and displays the patient's heartbeat sound in a graph on the phone's screen. The idea is to make detecting abnormal heart sounds easier for doctors and even patients themselves by taking advantage of the technology they already use every day. Patients with chronic heart and respiratory disease often end up in the emergency room when they have an episode, sometimes making several of the expensive and time-consuming trips every year. Data from the Eko DUO and Steth IO, hand-in-hand with artificial intelligence, could help doctors detect problems before they become emergencies and help patients stay healthier while decreasing their health-care costs.

Steth IO

Developments in drone technology are also set to disrupt the way we deliver care and advance the decentralization agenda in health care. Every year, cardiac arrests trigger the abrupt loss of heart function for 350,000 Americans, making it the leading cause of natural death in the country according to the American Heart Association. Today's drone manufacturers are taking steps to reduce this grim statistic. With each minute that passes, following the onset of symptoms, the chances of survival decrease by 10 percent, meaning that help often arrives too late. Starting in 2018, drone-maker Flirtey and US–based ambulance service Regional Emergency Medical Services Authority (REMSA) aim to start a trend that could save thousands of lives, using drones to deliver automated external defibrillators (AED) to victims.

In late 2016, Zipline, a San Francisco Bay Area–based robotics startup, set up distribution centers in Rwanda where its drones had made more than 1,400 flights carrying on-demand blood and emergency supplies over sixty-two thousand miles as of last fall. This year, the company will expand its medical delivery operations by launching a second base in Rwanda and new service in a larger neighboring country, Tanzania.

Swiss Post also launched a medical transport network in Lugano, Switzerland, using drones made by another Bay Area company, Matternet. So far the drones have made 350 deliveries, about five to fifteen per day. Beyond blood and medical supply deliveries, drones could transform another key component of health care—lab tests. Timely test results help doctors diagnose infections and reduce guesswork in prescribing medications. Some of those decisions have life-or-death implications and drone delivery of biological samples can dramatically reduce delivery times. In the same way, a lot of the technologies described in this book can easily be made accessible and fast to prosumers irrespective of where they are located.

Accelerating forces will continue to transform health care through cheaper, quicker, and more convenient care delivery. And as an investment in virtual health services continues to accelerate, breakthroughs will disrupt the market, creating new social interactions and experiences. New higher standards for service delivery will require organizations to shed the old ways of approaching health care and rapidly accept the digital era. When this happens, health care, as we know them, will cease to exist.

CHAPTER 5

Insurance to Outsurance

There is probably no more important an issue for the future of America than its long-term fiscal sustainability. And the current cost trajectory of health care in the USA is placing this in jeopardy. It is projected that health-care spending will, on average, rise 5.5 percent annually from 2017 to 2026 and will comprise 19.7 percent of the US economy in 2026, up from 17.9 percent in 2016. By 2026, health spending is projected to reach $5.7 trillion. The urgent need to tame runaway cost inflation in health care coupled with the advent of technology-driven health care that is democratizing data, empowering patients, and transforming care delivery and business models is resulting in an unprecedented revolution in health care.

Health insurers who happen to control the purse strings should be in the driver's seat in this transformation, but the sector overall has been slow to take control. To survive, let alone succeed in the era of health care 3.0, health insurers will need to capitalize on the catalysts of change out there—patients' elevation to prosumers driven by converging exponential technologies and the transition from a pipeline-based approach to care delivery to a platform-based one. With these shifts will come increased expectations and demands from insurers as well. Small, incremental changes will not suffice. Incumbent insurance companies looking to get ahead of the curve will need to fundamentally reinvent themselves to position themselves for success in the era of the new health economy or risk losing out to startups and nontraditional entrants who may be more responsive to the changing environment. There are already several examples of these, which we will discuss.

In this chapter, we explore how the insurance industry can and should transition from one where people pay in to mitigate against ill health to one where they get paid to stay healthy—from insurance to outsurance. In our proposed model, the insurance approach changes from a short-term transactional one where insurance premiums are determined by actuaries—a function of statistics and probabilities to a long-term relational approach in which the insurer and the insured partner improve behaviors and health outcomes. That's a 180-degree flip from being just in the business of quantifying and pricing risk to influencing and lowering risk. From a system where you need a pool of insured individuals, that is big enough to account for the variation in your predictive model to one where we have the data to be able to accurately understand and predict the risk of individuals.

Lessons from the Car Insurance Market

It's no surprise that the insurance industry in other sectors have already taken the lead in leveraging sensor-based technology, big data, and analytics to reward or pay their insured for good behavior. The car insurance industry is already being disrupted in a major way, but by incumbents. Progressive Insurance's Snapshot Automotive Sensor package (https://www.progressive.com/auto/discounts/snapshot/) personalizes your rate based on your actual driving. It's technically called usage-based insurance, and they are able to accurately monitor your driving by plugging in a sensor device into your car's on-board diagnostics port. That means you pay based on how and how much you drive instead of just traditional factors. It's simple. Drive safe and save. Drive extra safe and save even more. There are still other pricing factors, and your rate may increase with high-risk driving. But you're in control of what you pay for car insurance, and most drivers earn a discount that is tantamount to getting paid to drive safely. In fact, Snapshot rewards the average driver with a $130 discount.

Even scarier for the car insurance industry is the fact that self-driving cars will become the new normal, and with that, experts predict a sharp decline in accidents. Tesla predicts that its second-generation autopilot hardware will reduce the crash rate by 90 percent. Consequently, autonomous driving has the potential to save millions of lives and millions in repair costs around the globe every year and change the car insurance industry. In its North America, Hong Kong, and Australia markets, Tesla has already partnered with established insurers to match the insurance proportionate to the risk of the car. Their autosteer-eligible highway miles program discounts your insurance rate based on the number of miles you drive using their autopilot system. As self-driving technology improves and as adoption increases, the question will then be "Why would I need a car insurance policy if I never drive or if I don't own a car?" Accounting firm KPMG predicts that the motor insurance market may shrink by 60 percent by 2040. I suspect that this is a serious underestimate.

The key philosophy underpinning these strategic shifts in the car insurance industry is that the amount of information available to the insurer increases exponentially in this new data-centric, technology-enabled time, giving the company a much deeper understanding of the customer than has been possible so far. This allows them to more accurately predict risk at the individual level and to set premiums and rewards accordingly, or simply put, "Pay as you drive."

The Swedish went even further with the concept of paying out for compliance. In its efforts to get people to drive safely, the Swedish National Society for Road Safety in Stockholm also leveraged technology and gamification. Using existing traffic-camera and speed-capture technologies, a speed camera lottery device photographs all drivers passing beneath it. Each vehicle's speed is displayed to the drivers passing by and recorded by the system. Speedsters are photographed and issued a citation, with the proceeds going into a cash fund. Drivers who obey the speed law are also recorded and entered into the lottery, where they would be eligible to win some of the money from the speeders. The average speed of cars passing the camera dropped from thirty-two kilometer per hour to twenty-five kilometer per hour after. Now if only there was a way to pay car drivers to be polite.

Disruptive Health Insurance Companies

Health care is no different. Instead of vehicle telematics, why can't we use wearables and other exponential technologies to provide us with the data points that will inform premium pricing and make "pay as you live" insurance a reality? If behavior is a major contributor to the increase in preventable disease, then shouldn't health insurers play a role in helping their customers reduce the risk of preventable diseases? Like the car insurance industry, shouldn't they create products that proactively encourage and reward healthier living? The health industry will be far better off with a system that rewards or pays us to stay healthy (outsurance) as opposed to one where we pay in (insurance) to mitigate against illness and disease.

We are in the midst of a data deluge in health care. The spread of electronic medical records, the exponential sales growth of wearables to measure health parameters predicted to be around 250 million in 2019, coupled with growth of social networks, among others, is giving us unprecedented access to billions of data points that can now enable us to move toward a predictive, preventive, and personalized approached to care and the payment thereof, all enabled by AI and machine learning that has the ability to discover new correlations and long-term consequences.

Disruptive startups and companies are already starting to leverage exponential technologies such as digitization, artificial intelligence, genomics, sensors, and social networks to move toward an outsurance or "pay as you live" based model.

Lemonade (https://www.lemonade.com/) was the world's first peer-to-peer (P2P) insurance company. P2P reverses the traditional insurance model. They treat the premiums you pay as if it were your money. With P2P, everything becomes simple and transparent. Lemonade takes a flat fee, pays claims really fast, and gives back what's left to you or causes you care about. This makes insurance a social good rather than a necessary evil. Social networks will allow us to create true peer-to-peer insurance models as it enables to self-insure as a group with those that we can trust and vouch for. And when talking about life insurance, it's going to be difficult to ignore genomic data, especially as life insurance companies are exempt from Genetic Information Non–discrimination Act (GINA).

It's possible that soon groups with great genes will coalesce and self-insure. It's in their best interest to do so. You'll be able to upload your genomics data and find others in your peer group that have similar or better risk profiles than you do. For life insurance companies, this is a great alignment of incentives. These life insurance companies will use genomics information to help their clients stay alive longer. Why? Because the longer they are alive, the more premiums they can pay.

Oscar is a New York–based health insurance founded in 2012 and expected to have 260,000 members by 2018. Oscar offers all members free 24/7 telemedicine visits through their Doctor on Call service as a part of their membership, which is available both through the Oscar Android or iOS app or the Oscar web app. Their members can also can submit their Fitbit data, and if they reach their fitness goals, they get $1 every day in the form of Amazon gift cards. It helps keep people healthy and motivated with a simple but quantifiable reward.

Beam Technologies (https://www.beam.dental/) developed a Bluetooth-enabled sensor-based toothbrush that is paired with a smartphone application that monitors daily brushing. When you enroll in the Beam dental insurance plan, you receive a free toothbrush and dental supplies. They use a P2P approach and based on the "brushing score" (A to F) of the group, you can get up to a 16 percent discount or payback.

South African health insurer Discovery launched their Vitality (https://www.discovery.co.za/vitality/how-vitality-works) program several years ago and was probably the world's first "outsurance" product. Discovery Vitality rewards you for living well by encouraging you to exercise regularly, eat well, and do relevant health checks. They use a tiered reward system that only allows you to progress with a combination of physical activity, eating well, and regular health screenings. They also offer up to 25 percent cash back on thousands of healthy food items at partner stores; up to 25 percent cash back on preventative care and everyday personal and family care items at selected pharmacies and providers; up to 25 percent cash back on fitness devices, sports gear, and equipment; and up to 100 percent cash back on your gym fees when you achieve your monthly goals that are all tracked by a free (on condition you meet goals) Apple watch. Depending on your tier status, rewards can also be used for a host of other perks, including discounted flights, hotel accommodation, and meals.

Converging exponential technologies coupled with the big data that comes with it and the advances in AI and machine learning are going to dramatically disrupt and revolutionize the health insurance sector, and incumbents are going to be particularly vulnerable. To remain relevant, they will need to reinvent their business models in fundamental ways to get closer to customers, better understand their behaviors and risks, use data and technologies in new ways, and more. Most importantly, the health insurer of the future will need to be in a very different core business. So far, insurers have been in the business of pricing and underwriting risk. Risk has been static and, unfortunately, quantified using relatively little information. With this new approach, the insurer would leapfrog over the competition and enter a very different business, not just pricing and underwriting risk but influencing and reducing risk as well and doing so with a much better understanding of customers' behaviors and risk factors.

Futures Markets in Health Care

The increasing popularity and growth of health savings accounts (HSAs) will also lead to a new market structure—a health-care futures market—as an alternative to and enhancement of existing funding mechanisms. In other sectors of the economy, including agriculture and energy, there exists a robust futures-trading market where industry participants can transfer risk and achieve predictable pricing. Risk transference and mitigation are essential to financial and operating cost management. These critically important financial tools, however, have been nonexistent in the health-care sector. The agriculture industry, which has a long history of mostly stable pricing, has had a futures market since the 1830s. The energy industry saw difficult price spikes in the 1970s. In 1983, a futures market was created for oil and then for other energy commodities. While there have been a few price spikes since then, oil and gas prices have mostly been steady and affordable for the past thirty years. Health care faces a crisis today, similar to that in oil in the 1980s. Prices continue to spiral upward, and the aging population represents a surge in demand, and there are no capital market mechanisms in place to mitigate risk.

A futures contract is a legal agreement to buy or sell a particular commodity or financial instrument at a predetermined price at a specific time in the future. There would be contracts that are based on the cost of treatment (e.g., hip replacement) just like energy (oil, gas) or agriculture (wheat, corn). The futures market has been important in other industries because companies can buy or sell futures to ensure certainty in the future. Jet Blue has the opportunity to lock in its cost of fuel. Ford Motor Company, on the other hand, is pricing its new automobiles now and knows their cost of goods sold, except the health-care component that is estimated at $2,000 per car. A review of large- and medium-sized companies that are typically self-insuring themselves show very comprehensive hedging strategies to deal with interest rate and currency risk. Yet their health-care risk is, by and large, not hedged at all.

Even individuals who are facing higher deductibles and higher premiums don't have an effective way to manage the financial costs of their health care. In fact, 62 percent of all personal bankruptcies are due to medical expenses. The ability to mitigate this problem through new and better financial tools could provide real benefit to both consumers and producers of health-care services.

In health care, the "unit" will be the cost of treating one patient for one disease (one hip replacement). So based on historical data, Ford (or individuals using their HSAs for that matter) can purchase a projected number of units at a predetermined price (spot price) to be delivered at a specific time in the future. In this way, providers can sell their excess capacity in advance, and purchasers are able to lock in their expenses for specific health-care units. If Ford successfully implements employee wellness programs and their employees never need hip replacements, they could go back to a secondary market to resell units not used. This will apply to individuals as well. In this way, people will be incentivized to stay healthy so that they can benefit from gains by reselling health units, in effect making health an asset. A health-care futures market is a bold innovation capable of significant impact, so stay tuned to see how it develops. Stay tuned.